Research for Policy

Studies by the Netherlands Scientific Council for Government Policy

This series publishes internationally relevant studies of the Netherlands Scientific Council for Government Policy. Many of the cross-cutting issues that affect Dutch policymaking, also challenge other Western countries or international bodies. By publishing these studies in this international open access scientific series, the analyses and insights of The Netherlands Scientific Council for Government Policy (WRR) can contribute to the policy debate in other countries.

The Netherlands Scientific Council for Government Policy (WRR) is an independent strategic advisory body for government policy in the Netherlands. It advises the Dutch government and Parliament on long-term strategic issues that are of great importance for society. The WRR provides science-based advice aimed at: opening up new perspectives and directions, changing problem definitions, setting new policy goals, investigating new resources for problem-solving and enriching the public debate.

The studies of the WRR do not focus on one particular policy area, but on cross-cutting issues that affect future policymaking in multiple domains. A long-term perspective complements the day-to-day policymaking, which often concentrates on issues that dominate today's policy agenda.

The WRR consists of a Council and an academic staff who work closely together in multidisciplinary project teams. Council members are appointed by the King, and hold academic chairs at universities, currently in fields as diverse as economics, sociology, law, public administration and governance, health and water management. The WRR determines its own work program, as well as the content of its publications. All its work is externally reviewed before publication.

Ramy El-Dardiry · Paul 't Hart · Victor Toom · Annick de Vries · Suzanne Hulscher

People and Climate

How Social Infrastructure Powers Adaptation

WRR

THE NETHERLANDS SCIENTIFIC COUNCIL FOR GOVERNMENT POLICY

Springer

Ramy El-Dardiry
The Netherlands Scientific Council for
Government Policy (WRR)
The Hague, The Netherlands

Paul 't Hart
The Netherlands Scientific Council for
Government Policy (WRR)
The Hague, The Netherlands

Victor Toom
The Netherlands Scientific Council for
Government Policy (WRR)
The Hague, The Netherlands

Annick de Vries
The Netherlands Scientific Council for
Government Policy (WRR)
The Hague, The Netherlands

Suzanne Hulscher
The Netherlands Scientific Council for
Government Policy (WRR)
The Hague, The Netherlands

ISSN 2662-3684 ISSN 2662-3692 (electronic)
Research for Policy
ISBN 978-3-032-25830-4 ISBN 978-3-032-25831-1 (eBook)
https://doi.org/10.1007/978-3-032-25831-1

This work was supported by Netherlands Scientific Council for Government Policy.

The original submitted manuscript has been translated into English. The translation was done using artificial intelligence. A subsequent revision was performed by the author(s) to further refine the work and to ensure that the translation is appropriate concerning content and scientific correctness. It may, however, read stylistically different from a conventional translation.

Translation from the Dutch language edition: "Mens en Klimaat" by Ramy El-Dardiry et al., © Wetenschappelijke Raad voor het Regeringsbeleid 2025. Published by Wetenschappelijke Raad voor het Regeringsbeleid. All Rights Reserved.

This Springer imprint is published by the registered company Springer Nature Switzerland AG
The registered company address is: Gewerbestrasse 11, 6330 Cham, Switzerland

Preface

This book is a translation and light revision of the Dutch WRR report *Mens en klimaat. De kracht van sociale infrastructuur bij adaptatie*, which was published on 19 May 2025 and subsequently presented to the Dutch Minister of Infrastructure and Water Management. This translation was produced using an AI tool from Springer and was then edited by Jeroen van den Hark, a professional translator.

This publication was written by a project group consisting of Prof. Suzanne Hulscher (council member), Prof. Paul 't Hart (council member), Dr Ramy El-Dardiry (research fellow), Dr Victor Toom (research fellow) and Dr Annick de Vries (project coordinator). At an earlier stage of the project, Prof. Mark Bovens (former council member) and Elsenoor Wijlhuizen MSc (former research fellow) were also involved. Temporarily attached to the project group were Ellinore van Driel (intern), Pien Elferink (intern) and Alissa Stuut (intern). Pien Elferink was also involved in the process of translation. Dmitri Berkhout, Caroline Buser, Mitra Javanmardi and Magda de Wit provided invaluable support.

People and Climate. How Social Infrastructure Powers Adaptation is the product of an extensive study of the academic literature and policy documents. In addition, we conducted interviews with more than 60 external experts, including scientists, policymakers and administrators. Early on in this project, Prof. Arthur Petersen wrote the working paper *Klimaatkennis voor de lange termijn. Naar een adaptief klimaatbestuur*. This working paper is available in Dutch on the WRR website.

In the final phase of the project, the draft of the report was submitted for review to Prof. Godfried Engbersen (Erasmus University Rotterdam), Dr Menno van Duijn (Netherlands Institute for Public Safety) and Willem Ligtvoet (former programme leader at the PBL Netherlands Environmental Assessment Agency). Specific chapters were reviewed by Jos van Alphen (staff of the Delta Programme Commissioner), Prof. Linda Steg (University of Groningen and member of the Scientific Climate Council) and Dr Anne van Valkengoed (University of Groningen). We thank them for their comments and valuable suggestions.

The Hague, The Netherlands

Ramy El-Dardiry
Paul 't Hart
Victor Toom
Annick de Vries
Suzanne Hulscher

Competing Interests The authors have no competing interests to declare that are relevant to the content of this manuscript.

Contents

About the Authors

Dr. ir. Ramy El-Dardiry is a senior research fellow at the Netherlands Scientific Council for Government Policy (WRR). He studied applied physics (MSc) at the University of Twente and received his PhD from the University of Amsterdam. After spending several years in the private sector, he became a researcher at the Netherlands Bureau for Economic Policy Analysis (CPB) focusing on innovation and digitalization. Ramy is also an author of fiction. He published several short stories and a novel.

Prof. dr. Paul 't Hart is council member and vice chairman of the Netherlands Scientific Council for Government Policy (WRR). He is also professor of public administration at Utrecht University. His main research interests are public order and security, crises and crisis management, political and public sector leadership, political psychology, the relationships between politicians and public servants, government officials, learning from the successes and failures in public governance, public policymaking, organizations and partnerships, and Australian politics and governance.

Dr. Victor Toom is a research fellow at the Netherlands Scientific Council for Government Policy (WRR). He holds a PhD in the social sciences from the University of Amsterdam (2010). He held academic positions in England and Germany (2010–2018). He is an interdisciplinary scholar, with many publications in, amongst others, the British Journal of Criminology; Forensic Science International: Genetics; Science, Technology & Human Values; Security Dialogue; Studies in History and Philosophy of Biological and Biomedical Sciences; and Medical Anthropology.

Dr. Annick de Vries is senior research fellow at the Netherlands Scientific Council for Government Policy (WRR). She studied General Economics at the Erasmus University Rotterdam and obtained her PhD in Public Administration from the University of Twente. She then served as a consultant at Twynstra Gudde, where she advised public- and private-sector organisations. In 2013, she started working at the Rathenau Institute, where her portfolio included energy policy and energy transition. From 2016 to 2019, Annick worked at the Dutch Ministry of Economic Affairs and Climate Policy.

Prof. dr. Suzanne Hulscher is a council member of the Netherlands Scientific Council for Government Policy (WRR) since 2019. Also, she leads the cluster in Marine and Fluvial systems, department of Civil Engineering & Management at the University of Twente. She studied Theoretical Physics at Utrecht University and received her PhD at the Institute of Marine and Atmospheric Studies, Utrecht University on work partly done at Deltares. Her work mainly involves rivers, coasts and offshore areas. Herein she investigates physical processes as well as adaptation measures in these water systems to deal with climate change induced issues as sea-level rise, extreme weather patterns and more extreme river discharges. She (co-) authors over 200 journal papers and she has chaired this study as well as the WRR study on climate justice.

Chapter 1
Introduction: The Role of Social Factors in Climate Adaptation

Contents

Key Points

- Climate change is real and ongoing, confronting society with potentially disruptive events.
- The impact of these events is shaped in part by social factors. Focusing solely on physical adaptation measures is insufficient.
- In this book, we examine the interaction between the consequences of climate change and three social factors: how people deal with uncertainty, social cohesion and socio-economic position.
- We argue that climate resilience in the Netherlands will improve if the government also invests in social infrastructure, which enables people to look out for each other, increases mutual trust and offers perspectives for action.

1.1 A View on a Dutch Winter Landscape

In the Rijksmuseum in Amsterdam, there is a painting that many people immediately associate with the Netherlands: *Enjoying the Ice near a Town* by the Dutch painter Hendrick Avercamp. In this 17th-century work, we see young and old, men and

R. El-Dardiry et al., *People and Climate*, Research for Policy,
https://doi.org/10.1007/978-3-032-25831-1_1

Fig. 1.1 *Enjoying the Ice near a Town*, Hendrick Avercamp (c. 1620)[2]

women, enjoying themselves on the ice (Fig. 1.1). A diffuse winter light settles over the landscape; people are sledging, skating, chatting and playing. Viewed through a sociological lens, the painting reveals even more: all of society appears to come together on the ice. In the words of the museum: 'A beggar tries to scrounge a bit of money, and another man has cut a hole in the ice in the hope of catching fish. [...] A few well-to-do ladies wear velvet masks to protect the delicate skin around their eyes from the cold.'[1] For some people, a harsh winter exacerbated their dire circumstances, while for others, the ice brought joy. Avercamp seems to have made these social differences deliberately visible.

In our time, enjoying the ice is no longer a given. Climate change is causing Dutch winters to become milder, resulting in a decrease in the number of frost days. The last Elfstedentocht—a traditional skating tour of nearly 200 kms along 11 historic cities—was held almost 30 years ago, and skating on natural ice is no longer an annual occurrence.

Future generations may perhaps view the painting as an expression of bygone times. They will likely experience many winters with torrential rainfall and many summers marked by prolonged periods of drought and heat. In this book, we look at these changes in climate through a social lens—just like Avercamp did in his day.

Doing so exposes vulnerabilities in how a society prepares for a changing climate. In the painting, we see that protective measures against the cold were once reserved for a select group: only the wealthy ladies wear velvet masks and fur muffs. In our own time, the need to adapt to a new climate can similarly expose disparities. For example, measures to make homes, villages and cities climate-resilient require substantial investment. Not every member of society can afford these costs. The same goes for

[1] Rijksmuseum (n.d.)

[2] Avercamp (c. 1620).

some municipalities, housing associations and businesses. As a result, it is likely that the impacts of climate change will affect some much more severely than others.

A consideration of the social factors that play a role in climate adaptation offers new opportunities for society to become more resilient to climate change. Kampen—the city in the east of the Netherlands where Avercamp lived, worked and died—serves as an inspiration in this regard. Each year, a large group of volunteers gathers there to form the high-tide brigade. Together, they practise constructing a flood barrier to protect the city centre in the event the IJssel river were to overflow its banks. They cordon off the city, retrieve materials from storage depots and install flood defences where the old city wall once stood. Together, they keep Kampen safe.[3] Their collective action makes a positive difference—not only in terms of flood safety, but also because voluntary cooperation fosters a stronger sense of community.

Adaptation to climate change will require significant technical interventions, whether through reinforced dykes, tidal marshes or water-sensitive urban design. Fortunately, Dutch policymakers already pay considerable attention to these measures. But what do climate change and climate adaptation mean for *people*? And how can insights into human behaviour and social relations contribute to a more climate-resilient society?

In this book, we argue that climate resilience in the Netherlands will improve if the government also invests in social infrastructure, which enables people to look out for each other, increases mutual trust and offers perspectives for action. Our analysis focuses on three social factors: (1) how people deal with uncertainty, (2) the importance of social cohesion, and (3) the role of socio-economic position.

In this introduction, we briefly outline the anticipated consequences of climate change for the Netherlands (Sect. 1.2) and subsequently explain why social factors are important in climate adaptation policy (Sect. 1.3). Finally, we discuss the guiding principles (Sect. 1.4) and the structure of this book (Sect. 1.5).

1.2 The Consequences of Climate Change for the Netherlands

1.2.1 Changing Weather Patterns

Our climate is changing, and we are caught in the middle. The Royal Netherlands Meteorological Institute has found that the temperature in the Netherlands is now over 2 °C higher compared to the beginning of the twentieth century. In the Caribbean Netherlands—comprising the islands of Bonaire, Sint Eustatius and

---

[3] NPO (2023).

Saba—the temperature has increased by approximately 0.2 °C per decade since the 1980s.[4,5]

Scientists are gaining an increasingly detailed understanding of how higher temperatures are linked to observed weather phenomena and of which weather patterns we can expect in the future. All reports from the relevant research institutes and government agencies now unanimously state that the Netherlands is becoming warmer, wetter and drier, and sea levels will continue to rise. The most recent scenarios of the Royal Netherlands Meteorological Institute outline four possible climate pathways for the Netherlands around 2050, 2100 and 2150. These pathways 'are based on the amount of greenhouse gas emissions (and thus global warming) and the degree of changes in precipitation in the Netherlands.'[6] The Royal Netherlands Meteorological Institute did not assign probabilities to the different scenarios. The scenarios share many similarities in terms of what we can expect; the differences mainly concern the magnitude of the effects. In all scenarios, the meteorological institute anticipates more sunshine, an increase in average temperature and heat, longer periods of drought, more intense summer downpours, wetter winters and possibly stronger gusts of wind and downbursts during storms. Sea levels will continue to rise in any case. The Royal Netherlands Meteorological Institute has developed four comparable scenarios for the Caribbean Netherlands.[7] Temperatures and wind speeds will increase, while the average amount of precipitation will remain the same or decrease. In particular, Sint Eustatius and Saba must take into account a greater likelihood of severe hurricanes with heavy rainfall.[8]

In this book, we take these possible weather patterns as a starting point. We use them to understand and outline the societal consequences of climate change. That said, the actual future weather patterns may be different. For example, some scientists have found indications that the warm Gulf Stream in the Northern Hemisphere is coming to a halt. A weaker Gulf Stream would mean that temperatures in western Europe would not rise, but instead fall significantly.[9] We will return to this in Chapter 2. For now, we note that a better understanding of the interaction between social factors and the consequences of climate change—the central theme of this book—is still important, even if climate change were to unfold differently from the four scenarios outlined.

---

[4] The Kingdom of the Netherlands comprises a total of four countries: the Netherlands (including the Caribbean Netherlands, consisting of the special municipalities of Bonaire, Sint Eustatius and Saba), Aruba, Curaçao and Sint Maarten.

[5] KNMI (2023).

[6] KNMI (2023).

[7] The KNMI'23 climate scenarios for the Dutch Caribbean focus on the special municipalities of Sint Eustatius, Saba and Bonaire. The constituent countries of Curaçao, Aruba and Sint Maarten are not included in these scenarios. The WRR is preparing a publication that examines the themes of this book in relation to the Dutch Caribbean.

[8] KNMI (2023).

[9] Van Westen et al. (2024); Keulemans (2024).

1.2.2 *From Changing Weather Patterns to Climate Threats*

A changing climate has consequences for society as a whole. The European Environment Agency (EEA) emphasises that the aforementioned weather patterns will lead to increasing risks for the inhabitants of Europe.[10] When climate threats become a reality—in the form of 'climate events'; see also the glossary in Box 1.1—this can have significant social consequences. For example, heavy rainfall can cause severe local flooding, while extremely high temperatures have adverse effects on public health. Meanwhile, farmers are counting on reduced yields as a result of the increasing likelihood of extreme weather.[11] Moreover, climate events can occur simultaneously or in quick succession, thereby amplifying each other.[12] A combination of heat and drought, for instance, increases the risk of wildfires.

These climate threats have now become part of our reality. The overview of recent climate events in Infographic 1.1 further underscores this point. The 2019 heatwave—in which temperatures soared above 40 °C—resulted in 400 additional deaths in the Netherlands.[13] An international team of researchers demonstrated that the heavy rainfall which caused the flooding of the Geul river and the ensuing damage in the southern part of the province of Limburg in July 2021 cannot be separated from man-made climate change.[14] This same event also illustrates that altered weather patterns in other countries affect Dutch safety. The province of Limburg borders Belgium and France, and the water level of the Meuse river is largely determined by rainfall in these countries.

1.2.3 *Risks and Adaptation Policy*

In this context of changing weather patterns and increasing severity of climate events, the Netherlands is developing climate adaptation policy. This is defined as 'the process of adjusting to current or future climate effects in order to limit or prevent damage and to seize opportunities.'[15] In this book, we adopt a broad perspective on adaptation policy. In our view, it covers measures taken in anticipation of threats, disaster management and recovery policy, all of which aim to limit the damage caused by climate events.[16]

---

[10] EEA (2024).

[11] PBL (2024); NOS (2024a, 2024b).

[12] PBL (2024). In disaster research, these are referred to as compound disasters; see Zscheischler et al. (2018).

[13] CBS (2019). In the near future, significantly higher numbers of 'heat-related deaths' are to be expected in Europe; see García-León et al. (2024).

[14] Tradowsky et al. (2023).

[15] PBL (2024: 153); IPCC (2022: Annex II).

[16] In the earlier WRR publication Justice in Climate Policy: On the Distribution of Climate Costs (Springer Nature, 2024), we used the term 'climate policy'. In that publication, climate policy

Risk assessment is a key component of adaptation policy. These risks are generally defined as the product of likelihood and impact.[17] Climate change increases the *likelihood* of climate events. We must anticipate more periods of extreme weather. But what about the *impact* of such events? One way to gain insight into this is by estimating damage and casualties, as in the climate risk analysis conducted by the PBL Netherlands Environmental Assessment Agency.[18] In this analysis, the impact of climate risks is determined based on their effects on the economy, ecology and cultural heritage, as well as the potential number of fatalities and injuries. In this way, the PBL Netherlands Environmental Assessment Agency has demonstrated that heat has a significant impact due to the high number of human casualties, and that prolonged drought leads to substantial damage to cultural heritage.[19]

Such risk analyses are important tools for identifying vulnerabilities in adaptation policy. On the basis of such analyses, both the PBL Netherlands Environmental Assessment Agency and the EEA conclude that current Dutch adaptation policy is not adequate in all areas. More action is required, particularly in terms of policies addressing heat, drought and extreme precipitation. Given the rapid pace of climate change, this action needs to be taken swiftly.[20]

Box 1.1 Climate adaptation: key concepts

In this book, we adhere to commonly accepted definitions of climate-related terms.[21] We highlight a few of these below.

By *climate threat*, we refer to the potential occurrence of a physical climate-related or weather-related phenomenon with the capacity to cause harm. Examples include a prolonged heatwave or increasingly frequent hurricanes in the Dutch Caribbean. The starting point of this book is the threat posed by the increasingly dry, wet and warm climate.

A *climate risk* is a function of the climate threat and the exposure and vulnerability to it.[22] A climate risk is typically assessed by multiplying the probability of a climate threat occurring by its potential consequences. These consequences are often determined by estimating potential damage and casualties.

referred to mitigation, adaptation and climate-related damage. Mitigation falls outside the scope of this book. The focus is on policies aimed at adaptation, crisis management and recovery. Collectively, we refer to these as 'adaptation policy'.

[17] This is also referred to as the classical risk approach; see WRR (2008).

[18] The PBL conducts this risk analysis periodically. The risk analysis has a long history, dating back to a 2005 report by the MNP.

[19] PBL (2024).

[20] PBL (2024).

[21] PBL (2023, 2024).

[22] See, among others, IPCC (2014); IPCC (2022).

Box 1.1 (continued)

When a climate threat actually materialises, we refer to it as a *climate event*. Such an event may occur suddenly or unfold gradually. Its consequences may be immediately apparent, or only become visible at a later stage.

By a *climate-resilient society*, we mean a society in which people can live safely and healthily despite a changing climate.[23] The concept of 'climate resilience' can also refer to other systems, such as the natural environment.[24]

Resilience refers to the capacity of a social system (family, organisation, community) to adapt to and recover from extraordinary, significant disruptions that are perceived and experienced by that social system.[25]

1.3 The Complication: Climate Events are Transforming Society

Although the risk analyses discussed above are indispensable for shaping adaptation policy, some of the consequences of climate change remain underrepresented in such analyses. Not all consequences can be expressed in terms of damage or numbers of casualties. Climate events also affect the way we live together as a society.

For example, climate events can lead to societal disruption. Such disruption occurs, for example, when flooding destroys vital infrastructure and affects people's livelihoods and patterns of daily life, as was the case when the Geul river burst its banks in the province of Limburg.[26] In a 2014 report, the PBL Netherlands Environmental Assessment Agency researchers defined societal disruption as follows:

> 'In cases of societal disruption, the continuity of society is at stake. The threat to societal continuity becomes apparent when society ceases to function or when fundamental trust in society is lost. Societal disruption therefore has both a physical and a socio-psychological component.'[27]

According to this definition, societal disruption thus comprises not only a physical but also a social element. Human behaviour has a significant impact on the damage caused by a climate event. Such behaviour is often determined by social factors. In addition to physical damage to people, buildings and infrastructure, a climate event can also harm a community through the loss of social cohesion, collective trauma and the erosion of trust in public authorities or democratic institutions.

---

23 See Ten Velden et al. (2022).

24 There are also other definitions of climate resilience in circulation; see Klimaat voor Ruimte (2006: 11).

25 Van Duin (2011: 49).

26 WRR (2019).

27 PBL (2014).

The Netherlands Institute for Public Safety has indicated that when resilience is lacking during crises, the risk of societal disruption increases, particularly when a relatively large group within a community perceives or fears the loss of essential services.[28] Flooding and flood damage have repeatedly caused disruption throughout Dutch history.[29] Scenario studies suggest that certain climate events may trigger social and political unrest. In the worst-case scenario, fear, a sense of injustice, anger or panic may prevail. Such emotions can contribute to divisions within society or between citizens and the government.[30] For example, a heatwave resulting in high mortality may provoke indignation and undermine trust in the authorities.[31] Devastating events, such as hurricanes in the Caribbean, can create opportunities for looting, as occurred on Sint Maarten following Hurricane Irma.[32]

The Analyst Network for National Security—a knowledge network of Dutch intelligence services and research institutes that analyse risks and threats to national security—examined in 2014 what could happen in the event of flooding in the Lopikerwaard and Krimpenerwaard, a low-lying area in the west of the Netherlands, surrounded by rivers. What if a major dyke were to fail? From a socio-psychological perspective, the network anticipates—among other things—problems arising from the fact that 'a large-scale flood [...] could conflict with the image of the Netherlands as being in control of water management'.[33] There is a risk that citizens will blame the government for such a flood, which could undermine their institutional trust.

Even outside the specific context of acute climate events such as floods and tornadoes, society is affected by the consequences of climate change. The impact of climate change has tangible effects on people, such as reduced concentration and declining academic performance during periods of high temperatures or the uncertainty caused by the devaluation of homes in flood-prone areas. Climate change can improve existing social relations, but also exacerbate inequalities. Adaptation measures themselves can also have societal effects. In Box 1.2, we discuss an example from Wales in which people lost access to basic services as a result of adaptation measures implemented by the local government.

---

[28] NIPV (2024: 33–38).

[29] PBL (2014); Jensen (2024).

[30] RIVM (2022).

[31] During the 2003 heatwave, for example, considerable dissatisfaction arose among French citizens; see Zylberman (2012).

[32] NCTV (2022); AIV (2020).

[33] RIVM et al. (2014: 53).

Box 1.2 Adaptation policy as a divisive issue: tensions in Fairbourne

The approximately 700 residents of the village of Fairbourne in Wales have been referred to as the first climate refugees of the United Kingdom.[34] In 2014, they were informed about the planned dismantling of their village in 2054. Although that moment is still decades away, clear effects are already observable. What is going on?

The village of Fairbourne is situated half a metre above sea level at a river estuary. A sea wall separates the village from the sea. Maintaining the sea wall is expensive. These costs, combined with the anticipated future rise in sea level, led the municipal council to decide to dismantle the village in 2054. All infrastructure for electricity, sewage and drinking water will be removed.

The municipal council's decision to implement a managed retreat immediately led to a decrease in property values. One resident commented:

> 'They told us we'd even [have to] pay to demolish our own house, but with no compensation. So where would we go, what would we do? Live in a tent, or a houseboat if they're going to flood [the village]? Because you've worked all your life for this, and you want to leave it to your family, or your kids who're growing up. And you can't. And all of a sudden, where's it gone? Your life's just gone. [House prices have dropped by as much as] 40 per cent.'[35]

The decline in property values makes it impossible for some to move, as purchasing a new home is simply unaffordable. There are other widespread concerns among current residents, particularly because a well-considered plan is lacking. These include the disappearance of various services—from shops to health care—and the absence of an influx of young people.[36]

For another group, the drop in prices actually presents opportunities. A quarter of the homes have now been purchased as holiday homes, often for rental purposes.[37] The close-knit village has partly become a holiday village. As a result of these developments, community spirit and social cohesion are under threat.[38]

This example from Wales clearly demonstrates how profound the consequences of climate change and adaptation measures can be for residents, also in western Europe. Moreover, it highlights the importance of social factors. Because the residents experience little control over the situation, social trust and community cohesion are eroding.

It is essential to gain a thorough understanding of the social impact of climate change and of adaptation measures. Equally important is recognising the fact that social

[34] Wall (2019).

[35] 3Doc (2022).

[36] Arnal and Hilson (2023: 4).

[37] Forgrave et al. (2022); Wall (2019).

[38] Arnal and Hilson (2023: 4); Klinenberg (2002).

factors, in turn, influence the impact that a climate event may have. Individuals with a social network are more likely to reach safety or recover more quickly during a climate event than those without such support. Similarly, people with sufficient material resources at their disposal are better able to prepare for extreme weather events. Insight into social factors should therefore form part of the knowledge base underpinning adaptation policy. This requires policymakers to keep asking questions, such as:

- Are people aware of the risks they face?
- Do people feel they have sufficient capacity to act when it comes to managing these risks?
- Is there a recognition of the existence and causes of differences in vulnerability and resilience within society?
- What is required to prevent certain individuals or communities from being disproportionately affected?

To answer these questions, policymakers can draw on a growing body of knowledge.[39] Nevertheless, it is notable that adaptation policy typically relies on risk analyses that are largely confined to the physical domain. It is therefore unsurprising that solutions are also primarily sought within this physical domain, without taking into account the role of social factors. Technical interventions in the physical domain undoubtedly play an important part in climate adaptation policy and in creating a climate-resilient society, but on their own they are often insufficient. Physical measures and social processes interact with one another. In order to reduce climate vulnerabilities within society, solutions must also incorporate social factors. An approach that focuses too narrowly on physical measures alone may even unintentionally exacerbate existing vulnerabilities.[40] As this book will demonstrate, this scientific insight is also relevant to the Netherlands. The infrastructure that provides protection to the Dutch requires not only a physical, but also a social component.

1.4 The Importance of Social Factors in Adaptation Policy

With the book *Justice in Climate Policy* (2024), we have already drawn attention to the societal dimension of both adaptation and mitigation policies. We did so at the time from the perspective of a fair distribution of the costs of climate policy.[41] We

[39] The Analyst Network for National Security and the PBL are among several bodies that have published studies in which some of these questions are discussed; see RIVM et al. (2014: 158). In its most recent risk analysis, the PBL announced that it will address risk perceptions related to climate in greater detail in a forthcoming outlook; see PBL (2024).

[40] Hore et al., (2018: 152). In the literature, the term 'maladaptation' is also used in this context; see Schipper (2020).

[41] de Vries et al. (2024). The book Justice in Climate Policy is the English translation of the WRR-report *Rechtvaardigheid in klimaatbeleid* (2023).

are now doing so by examining the interplay between social factors on the one hand and the consequences of climate change on the other.

An adaptation policy that takes social factors into account contributes to societal resilience and thereby reduces the risk of disruption resulting from a climate event. There is a parallel with the coronavirus pandemic, during which the crisis threatened not only public health, but also social cohesion (see Box 1.3). At that time, the Netherlands proved to be ill-prepared for both. It now has only a limited window of opportunity to prevent a similar situation in the field of climate adaptation.

Box 1.3 Social factors during the coronavirus pandemic

At the end of 2019, the world was confronted with the outbreak of the coronavirus. Initially, the Dutch government attempted to keep the virus outside the national borders, among other things by introducing quarantine measures for travellers. These efforts proved unsuccessful. On 27 February 2020, the National Institute for Public Health and the Environment confirmed the first case of coronavirus infection in the Netherlands, and in March the country largely went into lockdown.[42] Intensive care units in hospitals became overcrowded. The virus resulted in thousands of deaths and caused long-term health complaints in a significant number of those infected.

Although the pandemic affected the entire country, there were significant differences in its societal impact. People without permanent employment contracts and newly established entrepreneurs suffered heavier financial blows than those in salaried employment and long-established businesses.[43] A much larger proportion of health care workers experienced severe long COVID symptoms compared to most other occupational groups, and the lockdowns affected the mental health of young people in particular.[44]

After an initial period of widespread solidarity throughout society, the pandemic also led to deep fractures in social cohesion. People became sharply divided in their assessments of the severity of the virus and the acceptability and effectiveness of the measures taken by the authorities to curb its spread. During the subsequent lockdowns and following the start of the vaccination campaign, various groups within society found themselves in opposition to one another. During the second address by the then Dutch Prime Minister Rutte on 14 December 2020, the jeering of demonstrators could be heard in the background. Partly as a result of this societal division over the policies pursued, trust in public institutions has suffered 'a significant blow'.[45] The consequences of

[42] Government of the Netherlands (2020).

[43] CPB (2022).

[44] NJI (2023).

[45] Engbersen et al. (2021); Engbersen (2022: 172); Snel et al. (2023).

Box 1.3 (continued)
this are still being felt, for example in a further decline in the vaccination rates for whooping cough and measles. In an increasing number of cities, vaccination rates are falling to such an extent that herd immunity is at risk.[46]

Interventions in the field of climate adaptation must therefore always address the interaction between physical measures and social processes. We elaborate this standpoint for three social factors derived from social-psychological, sociological and socio-economic research. The social science focus of this study is part of a rich tradition.[47]

The first factor concerns *how people deal with uncertainty*. World views, values and collective memory influence the way individuals and communities perceive climate risks and their capacity to act. This also affects whether people prepare for climate threats or not. Past climate events or disasters can also shape attitudes towards risks and perspectives for action.

The second factor concerns *social cohesion*: the degree of connectedness within a society. High social cohesion can encourage people to seek each other out and can be beneficial when a climate event occurs. Social cohesion also plays a role in recovery.[48] At the same time, the course of a climate event can also influence social cohesion within communities, both positively and negatively. The relationships between people and institutions also play a role.[49] If government policy leads to diminished public trust in the authorities, social cohesion will deteriorate.

Socio-economic position constitutes the third factor. Put bluntly: the wealthier one is, the safer one is.[50] Conversely, the poorer one is, the more severe the consequences of climate change can be. For example, differences in the location and quality of housing have a significant impact on how climate change affects various population groups. The same applies to individuals' physical health or the types of jobs they hold. A climate event can exacerbate social inequalities further through lingering effects, for instance in terms of educational attainment or debt problems.

These three factors each influence the extent to which individuals and communities are affected by climate change, as well as their capacity to cope with its consequences. In this book, we discuss how this works in practice and what knowledge we have obtained about this through the aforementioned fields of research.

As an aside, these social factors do not operate in separation from one another. For example, individuals in a socio-economically disadvantaged or socially isolated position often deal with uncertainty differently from those who have sufficient resources and an extensive network. Trust in the government and politics is also linked to

[46] NOS (2024c).

[47] Erikson (1995); Beck et al. (1992); Klinenberg (2002).

[48] Townshend et al. (2014).

[49] SCP (2024).

[50] See also Wildavsky (1988).

socio-economic position.[51] The fact that the consequences of climate change and adaptation measures differ across social groups poses risks to society. For instance, if vulnerable households are severely affected, it is conceivable that the individuals making up these households may turn away from the rest of society or from the government. Such social processes can, in the long term, undermine the legitimacy of policy.

This book does not aim to identify and discuss all relevant social factors and therefore does not provide a comprehensive societal risk analysis. Instead, our objective is to understand why focusing on social factors adds value and how this can contribute to a more robust Dutch adaptation policy. In addition to the three factors we address, there are other social factors that merit attention and are still scarcely discussed in the current discourse. These include the economic opportunities and threats associated with climate adaptation, or the importance of behaviour in the spread of new infectious diseases related to climate change.[52] How specific sectors, such as banking and insurance, respond to climate threats falls outside the scope of this report.

1.5 Structure of This Book

This book consists of six chapters.

In Chap. 2, we discuss current Dutch adaptation policy in greater detail. Using three key questions, we illustrate the main focus areas of this policy. Particular attention is given to the various types of measures that the government has implemented or may implement.

Chaps. 3, 4 and 5 address the three social factors described in the previous section. In Chapter 3, we discuss sociological and socio-psychological insights regarding the management of uncertainty. What risks and perspectives for action are there for individuals and communities? Chap. 4 examines the relationship between social cohesion and the consequences of climate change, while Chap. 5 addresses this in relation to socio-economic position.

In each of these chapters, we combine insights from existing research with knowledge obtained through interviews. To this end, we spoke with Dutch experts and policymakers. See the appendix for an overview of the interviewees.

In Chap. 6, we take stock and draw conclusions regarding the role of social factors in adaptation policy. We conclude with recommendations that enable policymakers to harness the potential of social factors within adaptation policy.

---

[51] SCP (2023).

[52] The WKR and the Health Council of the Netherlands are preparing an advisory report on the health aspects of climate change.

Trends in a changing climate

Due to climate change, we are increasingly confronted with heat, drought and heavy rainfall. In addition, sea levels are continuing to rise. Scientists, including those affiliated with World Weather Attribution, have demonstrated that many disasters in recent years are linked to climate change. Below, we present a number of these events.

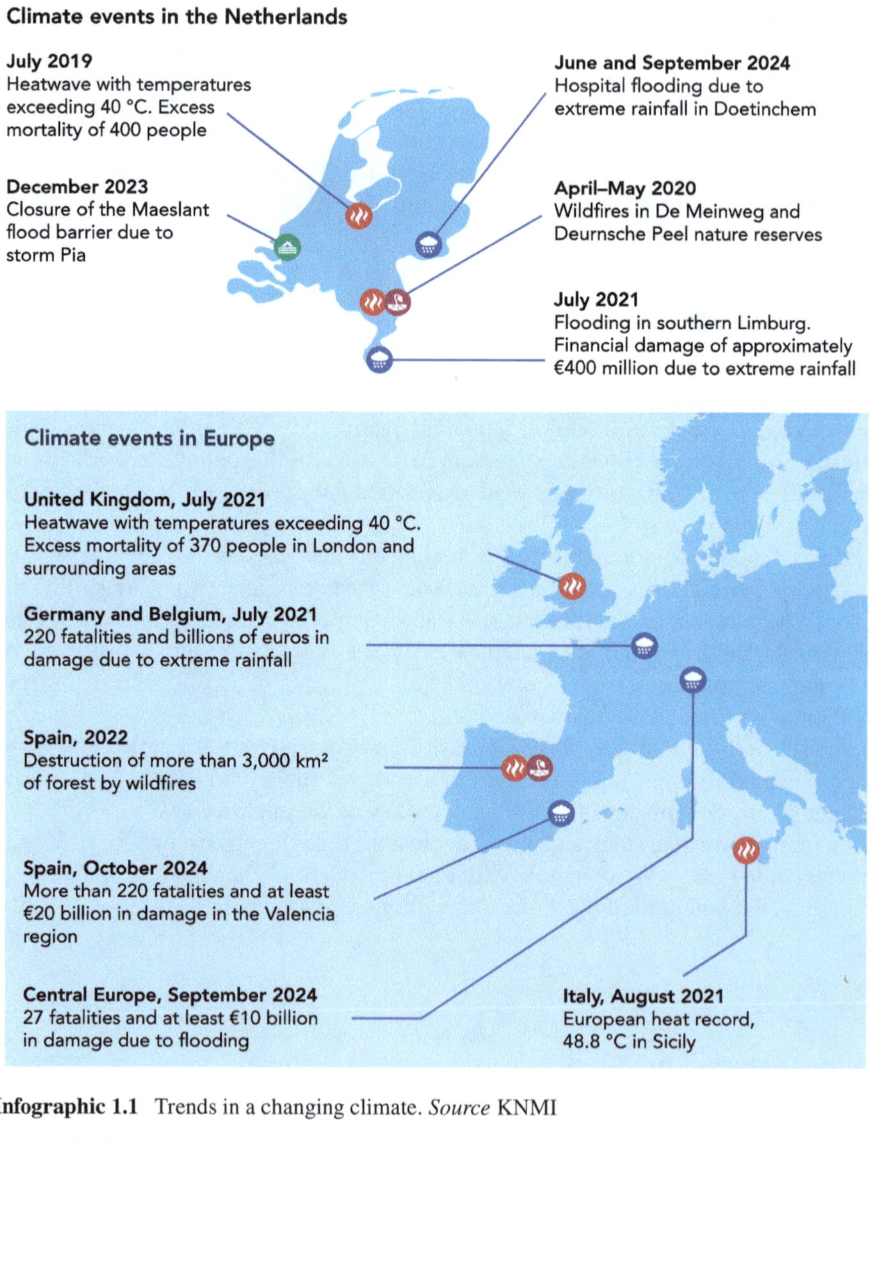

Infographic 1.1 Trends in a changing climate. *Source* KNMI

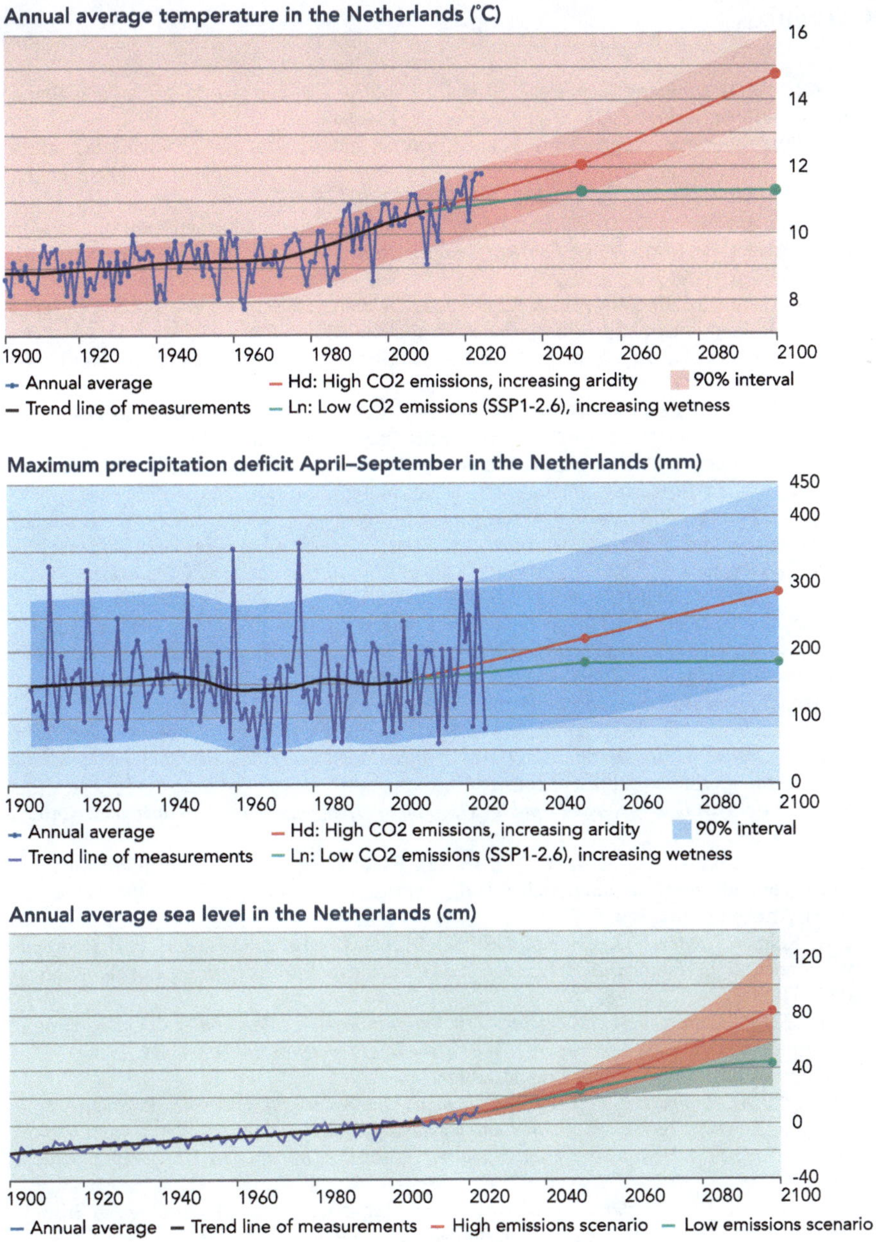

Infographic 1.1 (continued)

References

3Doc. (2022). *De klimaatverkenner.* Nederlandse Publieke Omroep (NPO). https://npo.nl/npo3/3doc-de-klimaatverkenner/28-10-2022/VPWON_1340943

AIV. (2020). *Veiligheid en rechtsorde in het Caribisch gebied.* Den Haag: Adviesraad Internationale Vraagstukken. https://pure.rug.nl/ws/portalfiles/portal/210770200/Veiligheid_en_rechtsorde_in_het_Caribisch_gebied_AIV_advies_116_202009.pdf

Arnall, A., & Hilson, C. (2023). Climate change imaginaries: Representing and contesting sea level rise in Fairbourne, North Wales. *Political Geography, 102*, 102839, 1–9. https://doi.org/10.1016/j.polgeo.2023.102839

Avercamp, H. (ca. 1620). *IJsvermaak bij een stad (Winterlandschap bij een stad)* [Olieverf op paneel]. Rijksmuseum. https://id.rijksmuseum.nl/200529588

Beck, J. V., Cole, K. D., Haji-Sheikh, A., & Litkouhl, B. (1992). *Heat conduction using Green's function.* Taylor & Francis.

CBS. (2019). *Hogere sterfte tijdens recente hittegolf*. Centraal Bureau voor de Statistiek. https://www.cbs.nl/nl-nl/nieuws/2019/32/hogere-sterfte-tijdens-recente-hittegolf

de Vries, A., Werner, G., Wijlhuizen, E., Toom, V., Bovens, M., & Hulscher, S. (2024). *Justice in climate policy: Distributing climate costs fairly*. Springer Nature.

EEA. (2024). *European Climate Risk Assessment*. https://www.eea.europa.eu/en/analysis/publications/european-climate-risk-assessment#:~:text=The%20first%20European%20Climate%20Risk%20Assessment%20%28EUCRA%29%20identifies,infrastructure%2C%20water%20resources%2C%20financial%20stability%2C%20and%20people%E2%80%99s%20health

Engbersen, G. (2022). De maatschappelijke impact van COVID-19—Sociologie van een pandemie. *De Sociologie en de Pandemie: Inzichten en Vooruitblik na Twee Jaar Coronacrisis*. https://pure.uvt.nl/ws/files/65910933/Sociologie_en_de_pandemie_compleet_2022_pdf

Engbersen, G., Van Bochove, M., de Boom, J., Bussemaker, J., El Farisi, B., Krouwel, A., Van Lindert, J., Rusinovic, K., Snel, E., & Van Heck, L. (2021). *De laag-vertrouwensamenleving. De maatschappelijke impact van COVID-19 in Amsterdam, Den Haag, Rotterdam & Nederland, vijfde meting*. Kenniswerkplaats Leefbare Wijken.

Erikson, K. (1995). *A new species of trouble: The human experience of modern disasters.* Norton & Company Inc.

Forgrave, R., Elliott, E. M., & Bain, D. J. (2022). Event scale hydrograph responses highlight impacts of widespread stream burial and urban infrastructure failures. *Hydrological Processes, 36*(5), Article e14584.

Hore, K., Kelman, I., Mercer, J., Gaillard, J. (2018). Climate Change and Disasters. In H. Rodríguez, W. Donner, & J. Trainor (Eds.), *Handbook of Disaster Research.* Springer. https://doi.org/10.1007/978-3-319-63254-4_8

IPCC. (2022). *Climate Change 2022: Impacts, Adaptation and Vulnerability. Working Group II Contribution to the Sixth Assessment Report of the Intergovernmental Panel on Climate Change.* C. U. Press. https://www.ipcc.ch/report/ar6/wg2/

Jensen, L. (2024). *Rampen: Een nieuwe geschiedenis van Nederland.* Uitgeverij Prometheus.

Klimaat voor Ruimte. (2006). *Naar een klimaatbestendig Nederland.* https://edepot.wur.nl/254919

Klinenberg, E. (2002). *Heat wave: A social autopsy of disaster in Chicago.* University of Chicago Press.

KNMI. (2023). *KNMI'23-klimaatscenario's.* Koninklijk Nederlands Meteorologisch Instituut. https://www.knmi.nl/kennis-en-datacentrum/achtergrond/knmi-23-klimaatscenario-s

NCTV. (2022). *Themarapportage klimaat en natuurrampen 2022.* Nationaal Coördinator Terrorismebestrijding en Veiligheid. https://www.nctv.nl/documenten/publicaties/2022/09/26/themarapportage-klimaat-en-natuurrampen-2022

NIPV. (2024). *Versterken van veerkracht. Naar een gezamenlijke aanpak van ongekende crises.* Nationaal Instituut Publieke Veiligheid. https://nipv.nl/wp-content/uploads/2024/05/20240501-NIPV-Crisisplan-Versterken-van-veerkracht.pdf

NJI. (2023). *Mentaal welbevinden van de jeugd: lessen uit de coronacrisis*. Nationaal Jeugdinstituut. https://www.nji.nl/publicaties/mentaal-welbevinden-van-de-jeugd-lessen-uit-de-coronacrisis

NOS. (2024a, May 3). *Bijna twee derde akkerbouwers zegt minder te verdienen door weersextremen*. Nederlandse Omroep Stichting. https://nos.nl/artikel/2519067-bijna-twee-derde-akkerbouwers-zegt-minder-te-verdienen-door-weersextremen

NOS. (2024b, August 20). *Extreem weer steeds groter probleem voor Nederlandse appels en peren*. Nederlandse Omroep Stichting. https://nos.nl/artikel/2533888-extreem-weer-steeds-groter-probleem-voor-nederlandse-appels-en-peren

NOS. (2024c, 14 March). *Vaccinatiegraad onder kinderen in Den Haag nog nooit zo gevaarlijk laag*. Nederlandse Omroep Stichting. https://nos.nl/artikel/2512681-vaccinatiegraad-onder-kinderen-in-den-haag-nog-nooit-zo-gevaarlijk-laag

NPO. (2023). *Langs de IJssel. De doorzetter*. Aflevering 1. Nederlandse Publieke Omroep. https://npo.nl/start/serie/langs-de-ijssel/seizoen-1/de-doorzetter

PBL. (2014). *Maatschappelijke ontwrichting en overstromingen*. Planbureau voor de Leefomgeving. https://www.pbl.nl/uploads/default/downloads/Maatschappelijke-ontwrichting-en-overstromingen.pdf

PBL. (2023). *Nationale klimaatrisicoanalyse 2022–2026* (5044). Planbureau voor de Leefomgeving. https://www.pbl.nl/uploads/default/downloads/pbl-2023-nationale-klimaatrisicoanalyse-2022-2026-uitwerking-methodiek-5044.pdf

PBL. (2024). *Klimaatrisico's in Nederland: De huidige stand van zaken* (5359). Planbureau voor de Leefomgeving. https://www.pbl.nl/system/files/document/2024-05/pbl-2024-klimaatrisicos-in-nederland-5359.pdf

Rijksoverheid. (n.d.). *February 2020: Eerste coronabesmetting in Nederland*. https://www.rijksoverheid.nl/onderwerpen/coronavirus-tijdlijn/februari-2020-eerste-coronabesmetting-in-nederland

RIVM. (2014). *Nationale Risicobeoordeling 6* (E/609042/14). Rijksinstituut voor Volksgezondheid en Milieu. https://www.rivm.nl/sites/default/files/2019-10/NRB%206%202014.pdf

RIVM. (2022). *Leidraad risicobeoordeling Rijksbrede Risicoanalyse Nationale Veiligheid*. Rijksinstituut voor Volksgezondheid en Milieu. https://www.rivm.nl/documenten/leidraad-risicobeoordeling-2022

Schipper, E. L. F. (2020). *Maladaptation: When Adaptation to Climate Change Goes Very Wrong*. One Earth. https://www.cell.com/one-earth/pdf/S2590-3322(20)30483-8.pdf

SCP. (2023). *Eigentijdse ongelijkheid: De postindustriële klassenstructuur op basis van vier typen kapitaal*. Sociaal en Cultureel Planbureau. https://www.scp.nl/publicaties/publicaties/2023/03/07/eigentijdse-ongelijkheid

SCP. (2024). *Factsheet Hoe staat het ervoor met de sociale cohesie in Nederland?* Sociaal en Cultureel Planbureau. https://www.scp.nl/publicaties/publicaties/2024/06/04/hoe-staat-ervoor-met-de-sociale-cohesie-in-nederland

Snel, E., Engbersen, G., de Boom, J., Seidler, Y., Sam, N., Enneking, G., l'Istelle, F., van Heck, L., & van Linder, J. (2023). *Corona & Cohesie: Dashboard Sociale Impact Corona. Verslag meting 5 (juni 2022) en meting 6 (november 2022)*. Risbo & Erasmus School of Social and Behavioural Sciences.

Ten Velden, C., Visser, M., & Van der Strate, E. (2022). NKWK-KBS *Monitoring Lokale Klimaatbestendigheid, Fase 2: Pilot indicatoren voor lokale klimaatbestendigheid op basis van bestaande datasets*. Deltares. https://www.deltares.nl/en/expertise/publications/rapportage-nkwk-kbs-inventarisatie-monitoring-lokale-klimaatbestendigheid-fase-1

Townshend, I., Awosoga, O., Kulig, J., & Fan, H. (2014). Social cohesion and resilience across communities that have experienced a disaster. *Natural Hazards, 76*(2), 913–938. https://doi.org/10.1007/s11069-014-1526-4

Tradowsky, J. S., Philip, S. Y., Kreienkamp, F., Kew, S. F., Lorenz, P., Arrighi, J., Bettmann, T., Caluwaerts, S., Chan, S. C., & De Cruz, L. (2023). Attribution of the heavy rainfall events leading to severe flooding in Western Europe during July 2021. *Climatic Change, 176*(7), 1–38.

Van Duin, M. (2011). *Veerkrachtige crisisbeheersing: nuchter over het bijzondere.* Nationaal Instituut Publieke Veiligheid. https://nipv.nl/wp-content/uploads/2023/01/20110201-NIFV-PA-Veerkrachtige-crisisbeheersing-nuchter-over-het-bijzondere-lectorale-rede-M-van-Duin.pdf
Van Westen, R. M., Kliphuis, M., & Dijkstra, H. A. (2024). Physics-based early warning signal shows that AMOC is on tipping course. *Science Advances, 10*(6), eadk1189. https://doi.org/10.1126/sciadv.adk1189
Wall, T. (2019, 18 mei). 'This is a wake-up call': the villagers who could be Brittain's first climate refugees. *The Guardian.* https://www.theguardian.com/environment/2019/may/18/this-is-a-wake-up-call-the-villagers-who-could-be-britains-first-climate-refugees
Wildavsky, A. (1988). Searching for safety. *Routledge.* https://doi.org/10.4324/9781351316248
WRR. (2008). *Onzekere veiligheid* (rapport nr.82). Wetenschappelijke Raad voor het Regeringsbeleid. https://www.wrr.nl/publicaties/rapporten/2008/10/01/onzekere-veiligheid
WRR. (2019). *Voorbereiden op digitale ontwrichting* (rapport nr.101). Wetenschappelijke Raad voor het Regeringsbeleid. https://www.wrr.nl/adviesprojecten/digitale-ontwrichting
WRR. (2023). *Rechtvaardigheid in klimaatbeleid: Over de verdeling van klimaatkosten* (rapport nr.106). Wetenschappelijke Raad voor het Regeringsbeleid. https://www.wrr.nl/adviesprojecten/klimaatbeleid/documenten/rapporten/2023/02/16/rechtvaardigheid-in-klimaatbeleid
Zscheischler, J., Westra, S., Van den Hurk, B. J. J. M., Seneviratne, S. I., Ward, P. J., Pitman, A., Agha Kouchak, A., Bresch, D. N., Leonard, M., Wahl, T., & Zhang, X. (2018). Future climate risk from compound events. *Nature Climate Change, 8*(6), 469–477. https://doi.org/10.1038/s41558-018-0156-3
Zylberman, P. (2012). Crises sanitaires, crises politiques. *Les Tribunes de la santé, 34*(1), 35–50. https://doi.org/10.3917/seve.034.0035

Chapter 2
Choices in Adaptation Policy

Contents

Key Points

- Climate change and its consequences transcend borders and affect society as a whole.
- In this chapter, we examine how Dutch adaptation policy is shaped by discussing three key questions. Which climate threats are the focus of attention? Do measures focus on preparedness, response, or recovery? And who is responsible?
- We conclude that adaptation policy is becoming increasingly important, that flood protection holds a dominant position and that policy often focuses on preventive measures in the physical domain.

2.1 The Societal Context of Adaptation Policy

The consequences of climate change are not confined by borders. An increase in average temperature leads to altered weather patterns across the globe. Wealthy countries such as the Netherlands are at risk of various climate events that can result in financial, social and ecological damage, as well as loss of life.[1] In addition, countries that emit relatively few greenhouse gases can also be severely affected. For many countries in the Global South, the risks are particularly significant.

[1] PBL (2024).

R. El-Dardiry et al., *People and Climate*, Research for Policy,
https://doi.org/10.1007/978-3-032-25831-1_2

Climate change is an issue that concerns society as a whole. For one thing, citizens, businesses and the authorities all contribute to greenhouse gas emissions, and global warming cannot be separated from our way of life. Moreover, the climate is 'everyone's concern'. The consequences of a changing climate are—to varying degrees—felt by all people, animals and ecosystems.[2]

Since 2015, nearly 200 countries and the European Union have signed the Paris Climate Agreement. With this agreement, the international community has set itself the goal of jointly containing the risks posed by climate threats,[3] although there has been a clear backlash against key commitments following the withdrawal of the United States from the agreement. The objectives of the Paris Agreement have been embedded in European and national climate legislation.[4] Article 2 of the Paris Agreement directly states that global efforts should focus on limiting further temperature increases (climate mitigation) and on enhancing adaptive capacity (climate adaptation). This dual objective makes sense: after all, climate mitigation and climate adaptation are inseparable. The less the international community succeeds in reducing greenhouse gas emissions, the greater the adaptation challenge becomes.[5] This is reflected in the scenarios of the Royal Netherlands Meteorological Institute discussed in Sect. 1.2: the extent to which greenhouse gas emissions are reduced determines the expected impact of climate events.[6]

Broad public support for climate policy depends on multiple factors.[7] The first relevant factor is whether people acknowledge that the climate is changing. If they do not, climate policy will—in the eyes of many citizens—seem like tilting at windmills. Second, it matters whether people view climate change as a positive or negative thing. This is because the consequences of climate change are not negative everywhere or in every respect. For example, rising temperatures may create new economic opportunities in some regions and economic sectors. The PBL Netherlands Environmental Assessment Agency has pointed out that climate change could have a positive impact on the cultivation of certain arable crops in the Netherlands.[8] Finally, it is relevant whether humans are seen as the main cause of climate change. For those who primarily regard climate change as a natural phenomenon, mitigation policy may seem of little use, although they may still recognise the importance of adaptation measures.

How these factors are viewed in the Netherlands has been studied extensively. A 2023 survey by Statistics Netherlands showed that 93% of Dutch people believe that the climate is changing. The survey also showed that about three-quarters of the

[2] IPCC (2022).

[3] The text of Article 2.1 reads: 'This Agreement, in enhancing the implementation of the Convention, including its objective, aims to strengthen the global response to the threat of climate change'; see UN 2015.

[4] Official Legislation Database (2019); Regulation (EU) (2021).

[5] See Climate Adaptation Knowledge Portal (n.d.); O'Brien (2009: 177).

[6] KNMI (2023).

[7] The three factors discussed here appear in several publications, e.g. Guy et al. (2014); Steg (2023).

[8] PBL (2015: 38).

population are concerned about this for future generations. Furthermore, six out of ten adult Dutch people see humans as the main cause of climate change. The vast majority of Dutch people believe that climate change can be partly (65%) or entirely (8%) prevented by human action.[9] These figures indicate that there is strong support in Dutch society for climate policy.[10] However, there are different perspectives on where policy priorities should lie—ranging from climate sceptics who would prefer not to allocate any resources to climate policy at all, to climate activists who view adaptation as an excuse to avoid taking drastic mitigation measures.

Nevertheless, both positions appear to be outliers. There is significant support for pursuing both mitigation and adaptation simultaneously. Studies in, among others, the United Kingdom and Switzerland show that the willingness to take mitigation measures is positively correlated with the extent to which people are prepared to take adaptation measures.[11] Environmentally conscious citizens often recognise the necessity of both mitigation and adaptation. Dutch research shows that acknowledging climate change is associated with support for adaptation policy.[12]

In this book, we demonstrate that social factors influence the impact of climate change and we discuss how social factors help determine the success of adaptation policy. In this chapter, we address three questions to illustrate the nature of Dutch adaptation policy as a prelude to that analysis. The first question concerns which climate threats are the focus of attention (Sect. 2.2). The second question concerns time: on which phases of climate events do adaptation measures focus (Sect. 2.3)? The third question concerns who is expected to take action (Sect. 2.4).

2.2 Question 1: Which Climate Threats are the Focus of Attention?

Climate scenarios from the Royal Netherlands Meteorological Institute show that both the Netherlands and the Dutch Caribbean are likely to face various types of climate threats. Not only is a new era beginning for the Dutch in their age-old struggle against rising water along the coast and rivers, but they will also have to learn to cope with heat, drought and more extreme weather events, such as intense summer downpours. In the Caribbean Netherlands, the risk of severe hurricanes is increasing.[13]

---

[9] At the same time, people are critical about the role the Netherlands should play in combating further climate change, with 33% indicating that the Netherlands should only implement strict climate policies if larger countries do so as well; see CBS (2023).

[10] SCP (2025).

[11] Brügger et al. (2015); Bateman and O'Connor (2016).

[12] Van Valkengoed et al. (2022).

[13] KNMI (2023).

The consequences of these different types of threats are diverse and potentially severe.[14] Prolonged drought impedes shipping on major rivers, leads to reduced crop yields or cooling water shortages for industry and power plants, and increases the risk of wildfires. The ecological damage from drought can also be very high, putting biodiversity under pressure.[15] Heat poses a similar problem for nature in the Netherlands and also leads to increased heat stress among people. Vulnerable groups, such as the elderly, people with chronic illnesses and young children, are particularly susceptible. Lastly, large amounts of rainfall in summer can not only cause flooding, but also lead to more waterborne infectious diseases.[16]

Citizens differ in the extent to which they recognise these threats. The aforementioned study by Statistics Netherlands shows that, in general, older people, more highly educated people, women and urban residents are more concerned about climate threats.[17] Concerns also vary depending on the type of threat. For example, young adults under 25 are more likely (32%) to be very concerned about a possible reduction in living space due to rising sea levels than people over 65 (23%).[18] The situation is reversed for heat: older people are clearly more concerned about this than younger people.[19]

With the National Climate Adaptation Strategy, introduced in 2016, the Dutch government sought to give greater direction to climate adaptation efforts.[20] This adaptation strategy will be updated in 2026. At the end of 2023, a significant recalibration took place with the publication of the National Climate Adaptation Implementation Programme. The aim of this implementation programme is explicitly to accelerate the process of making the Netherlands climate-resilient.[21] Partly as a result of a critical evaluation of the adaptation strategy, there is increasing attention to non-flood-related risks, such as heat and drought.[22] However, the PBL Netherlands Environmental Assessment Agency notes that the Netherlands still has considerable progress to make in these areas:

> For a number of climate risks, adaptation policy is already in place. Significant investments are being made and costs incurred to mitigate the effects of climate change. This is the case, for example, with flood protection. Policy for addressing climate risks arising from heat, drought and water nuisance due to extreme rainfall is often less well developed. The same

---

[14] PBL (2024).

[15] Witte et al. (2020).

[16] Limaheluw et al. (2023); Mulder et al. (2019).

[17] CBS 2023. A recent study by the Netherlands Institute for Social Research confirms this finding (SCP, 2025).

[18] CBS (2023).

[19] By way of illustration: among young adults aged up to 25, 55% report significant or some concern about summer heat, while this is 70% among those aged 75 and older; see CBS (2023).

[20] IenM (2016).

[21] IenW (2023).

[22] De Graaff et al. (2022).

> applies to combinations of climate risks that amplify each other's impact, as is the case for health effects. In many cases, policy in these areas is not yet sufficiently embedded.[23]

This observation is not unique.[24] Significant progress has been made in the field of flood protection. This is partly due to the establishment of the Delta Programme, a national programme with a long-term vision in which the Dutch government and other stakeholders cooperate on flood protection and freshwater supply. An independent government commissioner (the 'Delta Programme Commissioner') has been appointed to coordinate and drive this process, with access to a dedicated budget. This approach is embedded in law (see also Section 2.3). There is no comparable national, coordinated approach to other climate threats.[25]

Even less attention is paid to climate threats that are not covered in the Royal Netherlands Meteorological Institute scenarios.[26] Consider, for example, the risks associated with a weakening of the Gulf Stream. In such a scenario, the Netherlands would experience colder winters. However, such winters are not currently accounted for in policy documents. In Box 2.1, we discuss several potentially disruptive climate developments that generally fall outside the scope of models. Given the uncertainties inherent in model studies, it remains prudent to pursue an adaptation policy in which the type of threat is irrelevant, for example by focusing on enhancing societal resilience.

Box 2.1 Models and climate change

By definition, models are simplified representations of reality. Climate models are no exception. This simplification means that it is impossible to account for all natural phenomena. One example is volcanic eruptions. An increase in the frequency or magnitude of eruptions in the coming century could (temporarily) slow global warming. This is because particles released during eruptions generally reduce the amount of sunlight reaching the Earth's surface.[27]

In addition, global warming influences natural phenomena such as El Niño and La Niña. Recent scientific research indicates that the impact of both phenomena could increase under the influence of climate change.[28]

---

23 PBL 2024a: 134. Translation by WRR.

24 WKR (2023).

25 't Hart et al. (2023).

26 Nationaal Deltaprogramma (2024a).

27 Cole-Dai (2010). Underwater volcanic eruptions release large amounts of water vapour, which can lead to temperature increases; see Schoeberl et al. (2024).

28 Cai et al. (2018); Jo and Ham (2023); Geng et al. (2023).

Box 2.1 (continued)

Finally, scientists have identified tipping points. Small disturbances around these tipping points can trigger a transition to a different climate system on Earth.[29] Perhaps the best-known tipping point is the melting of ice sheets. The potential collapse of the West Antarctic ice sheet, for example, could lead to a more rapid rise in sea levels. However, there are many more tipping points of significance. The Atlantic meridional overturning circulation (AMOC, of which the Gulf Stream is a part), for instance, has a major impact on our climate. Climate change could contribute to a substantial weakening or even disappearance of the AMOC within this century. Such a change in the AMOC would have immense consequences for Western Europe. Instead of rising temperatures, it may then be faced with lower temperatures.[30] This tipping point appears to be closer than previously thought.[31] Another tipping point could occur if further global warming leads to the thawing of permafrost. Thawing permafrost will release additional greenhouse gases. This can trigger a self-reinforcing greenhouse effect that is not accounted for in most climate models.[32]

Models that directly inform Dutch adaptation measures also contain uncertainties. Many of these models use probabilities based on observations from the recent past. Extreme high river water levels are an example of this. To calculate the probability of high water, observations from the past century are used. Because the climate is changing, it is unclear how reliable such statistical models are for the future, even when climate change is taken into account. A high water level that is now expected once every 1,000 years may have a higher probability a decade from now.

The same applies to the frequency of extreme weather, resulting in flooding, heat stress and drought. One way to deal with these shifting expectations is to take potentially more extreme scenarios into consideration as well, without linking them to a specific recurrence interval. A good example of this are the model exercises conducted to examine what would happen if the large amount of rainfall that struck the southern province of Limburg in 2021 were to affect another area.[33]

[29] Lenton et al. (2008, 2009).

[30] Ditlevsen and Ditlevsen (2023); Van Westen et al. (2023).

[31] KNMI (2025).

[32] KNMI (2023).

[33] De Bruijn and Slager (2022).

2.3 Question 2: On which Phases of Climate Events do Adaptation Measures Focus?

Broadly speaking, three phases can be distinguished in dealing with acute climate events: before, during and after their occurrence. Each phase presents its own challenges and opportunities for measures to promote a climate-resilient society.

First, it is possible to take measures prior to a climate event to reduce its potential harmful impact. Coastal protection is a typical example, while greening cities to limit warming during heatwaves and reduce rainwater nuisance is another. In building design, preparatory measures might involve raising electrical outlets or including sunshades.

In addition, citizens, businesses and the authorities can take measures that take effect at the moment a climate event occurs. For example, drinking extra water during a heatwave can reduce the risk of heat stress, and stopping the watering of gardens and lawns can significantly reduce collective water use during drinking water shortages. Following a climate disaster, the government can declare a state of emergency, enabling decisive and far-reaching measures. This was the case during the floods in the Dutch province of Limburg in 2021. The mayor of Valkenburg, a severely affected town, issued an emergency ordinance to keep disaster tourists out of the area.[34]

The third phase takes place after a climate event. This phase is about recovery, although the term is somewhat misleading: reverting to the pre-disaster situation is usually not possible and not always desirable. Recovery policy involves not only reconstruction, but also transformation. For example, it may mean that damaged infrastructure and neighbourhoods are not necessarily restored but redesigned. A well-known example is the redesign of a residential area in Enschede, a city in the eastern part of the Netherlands.[35] In May 2000, a fire at a fireworks storage facility resulted in a series of devastating and deathly explosions, killing 23 people and injuring roughly 950 others. Hundreds of houses were completely destroyed or damaged. Recovery can also mean that affected residents and businesses are given the means to continue their lives and activities elsewhere or in a more climate-resilient manner.[36]

In the Netherlands, the focus is on preparatory measures. The most important preparatory measures relate to increasing flood protection. This is perhaps not surprising, as the Netherlands is a densely populated river delta. High water levels are potentially disastrous: nearly 60% of the country is susceptible to flooding and more than a quarter lies below sea level.[37] With the implementation of the Delta

[34] Municipality of Valkenburg (2021).

[35] Baart and Schaap (2007); De Graaf (2022).

[36] NIPV (2019).

[37] Ligtvoet and Knoop (2014).

Programme, a key pillar of the National Climate Adaptation Strategy, the Netherlands aims to remain a safe and liveable delta. See Box 2.2 for a brief overview of the Delta Programme.

Box 2.2 The Delta Programme

With the Delta Programme, the Dutch government aims to protect the Netherlands against high water and to ensure a sufficient supply of freshwater. The programme consists of three sub-programmes: Flood Risk Management, Freshwater and Spatial Adaptation.[38] Each sub-programme includes a 'Delta Decision', which sets out national frameworks, and a 'Delta Plan' with concrete measures for policy implementation.

The Flood Risk Management sub-programme aims to provide everyone who lives behind a dyke with a basic level of protection by 2050. Specifically, this means that the risk of death as a result of flooding may not exceed, on average, one in 100,000 years. To ensure this level of protection, the Delta Programme will initially focus on strengthening dykes. In total, 2,000 kms of dykes must be reinforced by 2050. As part of this major infrastructural task, nearly 219 kms had been reinforced by the end of 2023.

In addition, 400 hydraulic engineering structures, such as locks, pumping stations and storm surge barriers, must also be strengthened or replaced. By the end of 2023, work on 138 of these structures had been completed.[39]

The Freshwater sub-programme has the aim of making the Netherlands resilient to water shortages by 2050. In 2024, the PBL Netherlands Environmental Assessment Agency emphasised that drought constitutes a major climate threat.[40] The government employs a mix of policy instruments to address freshwater shortages. Notably, this sub-programme places greater emphasis on behaviour compared to the flood risk management programme. Efficient water use is an explicit sub-goal of this programme.

The aim of the Spatial Adaptation sub-programme is for the Netherlands to be climate-resilient and water-robust by 2050. This is to be achieved by reducing climate risks to health, the economy and safety resulting from heat, drought, water nuisance and flooding. Measures under the Spatial Adaptation sub-programme are largely the responsibility of municipalities, provinces, regional partnerships and water authorities.

A dominant focus on preparatory measures is not without risks. No matter how thorough the preparation, absolute safety can never be guaranteed. Taking into account the possibility of disasters can prevent significant social harm. Moreover, people may develop a false sense of security. For example, the Dutch believe that

[38] For the most recent Delta Programme, see Nationaal Deltaprogramma (2024b).

[39] Nationaal Deltaprogramma (2024b: 39).

[40] PBL (2024).

45% of the land area is vulnerable to flooding, while in reality this figure is nearly 60%.[41]

Scientists and policymakers are aware of this pitfall. The concept of 'multi-layer safety' focuses on a combination of measures before, during and after disasters (see Box 2.3). Multi-layer safety can also offer solutions for other climate threats, such as wildfires.[42] Although the concept has been embraced in the flood risk management domain, the Netherlands Court of Audit has found that policy implementation in this area still requires improvement. In practice, Dutch authorities remain primarily focused on developing preparatory measures.[43] In this book, we explicitly view adaptation policy as the full range of measures that the authorities can take to address the various phases of climate events.

Predominantly technical measures are important for enhancing climate resilience, but are not sufficient given the enormous challenges the Netherlands faces. The severity of a climate event is also determined by the social context in which it occurs. In order to improve climate resilience, social factors must be taken into account as well. How do citizens relate to climate change and how do they perceive associated risks, are citizens supportive of each other in times of crisis, and which types of households are most severely affected by climate events?

Box 2.3 Multi-layer safety

In 2009, the concept of 'multi-layer safety' was introduced in the National Water Plan, the policy framework that sets the general direction for water management.[44] The idea behind this concept is that policy focuses on 'protection against water and limitation of societal disruption in the event of an unforeseen disaster'.[45] To implement multi-layer safety, the Dutch government focuses on (1) prevention to avoid flooding, (2) limitation of the consequences of potential flooding through spatial planning, and (3) assurance of effective disaster response and crisis management.[46]

In a recent report, the Netherlands Court of Audit stated that multi-layer safety is difficult to implement in practice. According to the report, the Dutch government mainly focuses on strengthening dykes, while measures aimed at consequence mitigation and crisis management receive less attention. Insufficient cooperation with the safety regions is one of the reasons for this.[47]

---

[41] Van Thiel and Mol (2020).

[42] NIPV (2023).

[43] Netherlands Court of Audit (2023).

[44] The National Water Plan provides the policy framework for six years. The Delta Programme addresses implementation and the necessary cooperation agreements.

[45] VenW (2009: 71). Translation by WRR.

[46] VenW (2009: 71–75).

[47] Netherlands Court of Audit (2023).

Box 2.3 (continued)

Following the floods in the province of Limburg in 2021, experts argued that the concept of multi-layer safety needs to be expanded. Water awareness and self-reliance must be promoted, and there should be an additional layer to address recovery after the initial management of the crisis. These adjustments are deemed necessary because the limits of the Dutch water system have been reached. Citizens, businesses and the authorities must 'contribute more than before to reducing the impact of extreme weather events'.[48]

2.4 Question 3: Who is Responsible for Taking Adaptation Measures?

A climate-resilient society is achieved through the interplay of the government, businesses and citizens. When citizens and businesses take their own initiatives, the government primarily has a facilitating role to play.[49] However, although the Dutch recognise climate threats, they could do much more to cope with these threats than is currently the case. This raises the question of whether and how the government should encourage citizens and businesses to take action.

Various factors play a role when it comes to taking adaptive action. What does it cost? Are people in a position to ensure their own climate resilience? Do they feel that they truly have control over their own fate? Do they also feel responsible for the greater good?[50] In a study on citizen participation in climate adaptation, for example, it was found that a homeowners' association in Rotterdam mainly saw heat stress as a problem for residents of the upper floors of apartment buildings.[51] With regard to flood protection, 17% of Dutch people have a great deal of confidence and 64% have a moderate amount of confidence in the government to provide adequate measures.[52]

Some hazards are presented as a public matter, while other threats are considered a private issue. These differences are reflected in Dutch laws and regulations. The Water Act contains standards for the protection level of public water defences and describes the responsibilities of water authorities.[53] By contrast, the management of threats related to heat has been delegated to the private sector. For example, Dutch occupational health and safety law requires employers to ensure that workplace

---

[48] Rijksoverheid. (2023: 8).Translation by WRR.

[49] Van der Steen et al. (2016).

[50] Van Valkengoed and Steg (2019).

[51] Buijs et al., (2021: 75).

[52] Van Thiel and Mol (2020). SCP (2025).

[53] As of 1 January 2024, the new Environment and Planning Act is in force. This act replaces, among others, the Water Act (and 26 other laws concerning the physical living environment).

temperatures do not cause health risks for employees.[54] There are no public standards to prevent specific issues, such as the formation of urban heat islands. However, in 2021, the government introduced a heat standard for new buildings with the Structures (Living Environment) Decree.[55] With respect to heat, the government mainly fulfils an informative and warning function. The National Heat Plan primarily consists of informing local partners, such as the municipal public health services, health care professionals, informal caregivers and vulnerable groups.

This approach is a political choice. The question is whether this choice will hold if the Netherlands is confronted with an intense and prolonged heatwave that causes major problems in care institutions and significant excess mortality in cities. Such an experience could lead to the politicisation of the issue, nationalisation of the approach and the 'rollout' of standards, regulation and supervision.[56]

For other climate threats outside the domain of flood protection, the responsibility for taking measures also often lies with private parties or regional authorities. The division of responsibilities is not absolute. For example, water utilities are primarily responsible for ensuring an adequate supply of high-quality drinking water, but a government-mandated priority sequence prescribes which sectors are given priority during prolonged water shortages.

Notably, much of the government's adaptation policy does not have a mandatory character. For instance, the government has developed a national framework for the construction sector. This provides an overview of how to build in a climate-adaptive manner in the absence of national standards.[57] Similarly, when it comes to the planning assessment framework—a tool intended to help municipalities, water authorities and provinces select locations for new housing and buildings—no decision has yet been made regarding its formal adoption.[58] Nevertheless, the government plays an important role in enhancing climate resilience in the Netherlands. We distinguish three types of measures the government can take to increase the climate resilience of society.

First, the government can provide infrastructure. This can be achieved through technical interventions in the physical environment, as is done in the Delta Programme through the reinforcement of dykes and flood defences (see Box 2.2). These protection measures often have the characteristics of a public good. Without government involvement, such protection measures are unlikely to be realised. Involving citizens in participatory processes has become standard practice. Physical measures to reduce the impact of climate events can also be taken in areas behind dykes. For example, electricity substations can be given extra protection against high water by constructing them on stilts.[59]

---

[54] See Section 2.2 of the Water Act and Article 6.1 of the Working Conditions Decree.

[55] Article 4.149b of the Structures (Living Environment) Decree.

[56] Van Duin (1992); Birkland (2006); Helsloot and Scholtens (2015).

[57] Ministry of the BZK (2023).

[58] Storymaps (n.d.); Government of the Netherlands (2024).

[59] Netherlands Court of Audit (2023).

Working with nature is increasingly becoming a guiding principle in the development of infrastructure. For example, Dutch dykes can be reinforced with salt marshes on the seaward side, and mangrove forests can be established in the Caribbean. Both provide natural protection against flooding. The 'sand motor' at Kijkduin beach on the North Sea coast is an internationally renowned example where coastal protection goes hand in hand with nature development. This artificial sand peninsula, constructed in 2011, protects part of the North Sea coast by allowing sand to be distributed naturally by wind, waves and currents. This not only strengthens the coast, but also creates space for nature and recreation. In scientific literature, such measures are referred to as nature-based solutions or building with nature. Promoting greater biodiversity generally makes ecosystems more resistant to external disturbances and more resilient in coping with shocks. Greening the built environment naturally provides cooling and rainwater storage.

In addition, the government can provide non-physical forms of infrastructure, for example in the field of knowledge. Government-affiliated organisations play a role in the timely identification of threats and the provision of information. Examples include the climate scenarios published by the Royal Netherlands Meteorological Institute in 2023, the monitoring of climate-related infectious diseases and the issuing of weather warnings. In Chaps. 4 and 6, we will see that social infrastructure is essential to the implementation of adaptation policy as well.

The government can also choose to promote responsible behaviour. Legislation and regulation are established instruments for influencing the choices of businesses and citizens. The government can reduce the climate vulnerability of society through regulation, for example by no longer issuing building permits in certain flood-prone areas or by setting requirements for new buildings and gardens. Price incentives are another way to encourage citizens and businesses to change their behaviour.[60] Subsidies for installing insulation are an example of how the government seeks to make climate measures more attractive.

Education and public information are also part of the policy toolkit for changing behaviour. By giving more attention in agricultural education to the cultivation of climate-resilient crops, the likelihood increases that farmers will actually grow such crops. Similarly, architectural education can stimulate the construction of climate-resilient homes. Municipalities can use public information to show residents which climate risks their neighbourhoods face and what action they can take themselves to make their homes or gardens more climate-resilient.[61]

In the recovery phase following a climate event, the Dutch government may decide to assist households and businesses. In such cases, a recovery programme is established in which the government provides compensation, financial support or another form of restitution for material or immaterial damages suffered. Sometimes, the government chooses to go beyond what is legally required, for example when the damage is so extensive that it overwhelms many people. This is referred

[60] WRR (2023).

[61] See, for example, the climate maps of Amsterdam and Heemstede: Municipality of Amsterdam (n.d.); Municipality of Heemstede (n.d.).

to as ‘non-mandatory’ compensation for loss. The decision to compensate citizens and businesses in this way is ultimately a political one. Views on, for instance, the role of government and solidarity play an important part in such decisions. The Disasters (Compensation) Act is an example of the aforementioned non-mandatory compensation.

The third category of policy options concerns coordination. For example, the government can stimulate cooperation between private parties by bringing them into contact with one another. As an example, the government’s Action Programme for Climate Adaptation in Agriculture facilitates research and innovation.[62] The development of an adaptation strategy, for example through the National Climate Adaptation Strategy, is another example of government coordination. In the field of water management, there are several national coordination committees. Examples include the National Flood Threat Coordination Committee, which plays a role in early warning of increased flood risks, and the National Water Distribution Coordination Committee, which becomes active during prolonged periods of drought and imminent water shortages.[63]

If a climate event is so severe that it constitutes a disaster or crisis, the government focuses on disaster management. If an event extends beyond municipal boundaries, responsibility is transferred to the chair of the safety region in question. Each region has its own plans and protocols. For example, during the Limburg flood in 2021, a flood response plan was used.[64]

The above examples show that the Dutch government implements a wide range of measures to make society more climate-resilient. The choice of a particular instrument is by no means always set in stone. In the case of major risks, it is often more effective to invest in public, preventive infrastructure. A cost–benefit analysis for these often expensive measures is more likely to yield a positive outcome in such cases. Nevertheless, it is not only the result of a cost–benefit analysis that determines where and whether public resources are deployed to address climate threats. The framing of risks and decisions about the balance between individual responsibility and government intervention are political matters.[65]

2.5 Conclusion

In this chapter, we examined the context of climate adaptation policy. Combined with mitigation measures, adaptation measures must ensure that the impact of climate change remains within manageable limits. Whereas Dutch climate policy initially focused primarily on mitigation, adaptation has gained a more prominent role over the past decade.

---

[62] LNV (2020).

[63] Living Environment Information Centre (n.d.).

[64] North and South Limburg Safety Regions (2020).

[65] WRR (2008).

In the Netherlands, climate adaptation policy is closely intertwined with the centuries-old struggle against water. This history means that other climate threats, such as heat and drought, have received relatively less attention. Many measures, particularly in the field of flood protection, are aimed at using technical interventions to withstand the effects of climate change. Dyke reinforcement and the improvement of hydraulic engineering structures play a prominent role in this regard. The concept of 'multi-layer safety' is not yet established policy practice. In this book, we adopt a broad perspective on adaptation policy, which also includes response and recovery.

The following chapters will show that adaptation policy can be substantially improved by systematically incorporating social factors. We will look in particular at how three social factors—dealing with uncertainty, social cohesion and socio-economic position—influence societal resilience to climate events. In Chap. 6, we conclude in light of this analysis that existing Dutch adaptation policy (as discussed above) can be significantly strengthened by combining physical and social interventions.

References

Algemene Rekenkamer. (2023). *Voorbij de dijk. Keuzes in het waterveiligheidsbeleid.* https://www.rekenkamer.nl/publicaties/rapporten/2023/10/12/voorbij-de-dijk

Baart, T., & Schaap, T. (2007). *Bouwplaats Enschede: Een stad herschept zichzelf*. NAi Uitgevers.

Bateman, T. S., & O'Connor, K. (2016). Felt responsibility and climate engagement: Distinguishing adaptation from mitigation. *Global Environmental Change*, 41, 206–215. https://doi.org/10.1016/j.gloenvcha.2016.11.001

Birkland, T. A. (2006). *Lessons of disaster: Policy change after catastrophic events.* Georgetown University Press.

Brügger, A., Morton, T. A., & Dessai, S. (2015). Hand in Hand: Public Endorsement of Climate Change Mitigation and Adaptation. *PLOS ONE, 10*(4), e0124843. https://doi.org/10.1371/journal.pone.0124843

Buijs, J., Terpstra, T., Heuvel, J., Roest, A., Boogaard, F., Van der Maas, P., Fliervoet, J., Witteveen, L., Baan, F., & Heikoop, R. (2021). *Burgerparticipatie in klimaatadaptatie.* https://steenbreek.nl/wp-content/uploads/2022/01/Eindrapport-Burgerparticipatie-in-Klimaatadaptatie.pdf

BZK. (2023). *Aanbieding Landelijke maatlat voor een groene klimaatadaptieve gebouwde omgeving en verkenning groennorm.* Ministerie van Binnenlandse Zaken en Koninkrijkrelaties. https://open.overheid.nl/documenten/ronl-cba8e18abd991aba311834eb957f270aa3081c80/pdf

Cai, W., Wang, G., Dewitte, B., Wu, L., Santoso, A., Takahashi, K., Yang, Y., Carréric, A., & McPhaden, M. J. (2018). Increased variability of eastern Pacific El Niño under greenhouse warming. *Nature, 564*(7735), 201–206.

CBS. (2023). *Klimaatverandering en energietransitie: Opvattingen en gedrag van Nederlanders in 2023.* Centraal Bureau voor de Statistiek. https://longreads.cbs.nl/klimaatverandering-en-energietransitie-2023/

Cole-Dai, J. (2010). Volcanoes and climate. *WIREs Climate Change, 1*(6), 824–839. https://doi.org/10.1002/wcc.76

De Bruijn, K., & Slager, K. (2022). *Wat als ' de waterbom' elders in Nederland was gevallen?* Deltares. https://publications.deltares.nl/11206890_010_0006.pdf

De Graaf, K. (2022). *Naar een dynamische* (Ed.) milieuvergunning. Bye-bye bestaande rechten? In Milieurecht in transitie: Jubileumbundel 40 jaar VMR (pp. 87–95). Boom Juridisch.

Ditlevsen, P., & Ditlevsen, S. (2023). Warning of a forthcoming collapse of the Atlantic meridional overturning circulation. *Nature Communications, 14*(1), 1–12.
Gemeente Amsterdam. (n.d. a). *Klimaatadaptatie.* https://maps.amsterdam.nl/klimaatadaptatie/
Gemeente Heemstede. (n.d. b). *Klimaatkaarten.* https://www.heemstede.nl/klimaatkaart
Gemeente Valkenburg. (2021). Noodverordening. https://www.valkenburg.nl/data/downloadables/5/0/2/noodverordening-21-juli-2021.pdf
Geng, T., Jia, F., Cai, W., Wu, L., Gan, B., Jing, Z., Li, S., & McPhaden, M. J. (2023). Increased occurrences of consecutive La Niña events under global warming. *Nature, 619*(7971), 774–781.
Government of the Netherlands / Rijksoverheid. (2023). *Beleidstafel wateroverlast en hoogwater.* https://open.overheid.nl/documebr/469nten/dpc-327e4ad8c480deec85c2c5b625fa723b8554d0e5/pdf
Groeneveld, G., Kodde, J., Waning, S. (2024, 21 juli). *Extreme wateroverlast in Twente: drijvende auto's in Enschede,* A1 tussen Oldenzaal en Duitse grens urenlang afgesloten. Tubantia. https://www.tubantia.nl/enschede/extreme-wateroverlast-in-twen-te-drijvende-autos-in-enschede-a1-tussen-oldenzaal-en-duitse-grens-urenlang-afgesloten~a616b050/?referrer=https%3A%2F%2Fwww.google.com%2F
Guy, S., Kashima, Y., Walker, I., & O'Neill, S. (2014). Investigating the effects of knowledge and ideology on climate change beliefs. *European Journal of Social Psychology, 44*(5), 421–429. https://doi.org/10.1002/ejsp.2039
't Hart, P. t., Pot, W., & Biesbroek, G. (2023). Klimaatadaptatie: hoe leggen we onze oogkleppen af? In T. Overmans, M. Honingh, & M. Noordegraaf (Eds.), *Maatschappelijke bestuurskunde: Hoe verbindende bestuurskundigen (kunnen) inspelen op maatschappelijke vraagstukken* (pp. 29–50). Boom.
Helsloot, I., & Scholtens, A. (2015). *Krachten rond de risico-regelreflex beschreven en geïllustreerd in 27 voorbeelden.* Boom Juridische Uitgevers.
IenM. (2016). Nationale klimaatadaptatiestrategie (NAS). *Ministerie van Infrastructuur en Milieu.* https://klimaatadaptatienederland.nl/publish/pages/120542/nas_rapport_5_2-1-1.pdf
IenW. (2023). Nationaal Uitvoeringsprogramma Klimaatadaptatie Slimmer, intensiever, voor en door iedereen. Ministerie van Infrastructuur en Waterstaat. https://www.rijksoverheid.nl/documenten/rapporten/2023/11/17/bijlage-2-nationaal-uitvoeringsprogramma-klimaatadaptatie-nup-ka
Informatiepunt Leefomgeving. (n.d.). *Landelijke Coördinatiecommissies bij watercrises.* https://iplo.nl/thema/water/crisismanagement-water/landelijke-coordinatiecommissies/
IPCC. (2022). *Climate Change 2022: Impacts, Adaptation and Vulnerability.* (Working Group II Contribution to the Sixth Assessment Report of the Intergovernmental Panel on Climate Change). C. U. Press. https://www.ipcc.ch/report/ar6/wg2/
Jo, H.-S., & Ham, Y.-G. (2023). Enhanced joint impact of western hemispheric precursors increases extreme El Niño frequency under greenhouse warming. *Nature Communications, 14*(1), 6356.
Klimaatadaptatie Nederland. (n.d. a). *Impulsregeling klimaatadaptatie.* https://klimaatadaptatienederland.nl/beleid/nationale-aanpak/dpra/deltaplan/stimuleren-faciliteren/impulsregeling-klimaatadaptatie/
Klimaatadaptatie Nederland. (n.d. b). *Voorbeeldlijst maatregelen Impulsregeling klimaatadaptatie.* https://klimaatadaptatienederland.nl/publish/pages/179576/voorbeeldlijst_maatregelen_impulsregeling.pdf
KNMI. (2023). KNMI'23-klimaatscenario's. Koninklijk Nederlands Meteorologisch Instituut. https://www.knmi.nl/kennis-en-datacentrum/achtergrond/knmi-23-klimaatscenario-s
KNMI. (2025). *De staat van ons klimaat 2024. Koninklijk Nederlands Meteorologisch Instituut.* https://cdn.knmi.nl/system/data_center_publications/files/000/072/295/original/KNMI-publicatie_25-01.pdf
Lenton, T. M., Held, H., Kriegler, E., Hall, J. W., Lucht, W., Rahmstorf, S., & Schellnhuber, H. J. (2008). Tipping elements in the Earth's climate system. *Proceedings of the National Academy of Sciences, 105*(6), 1786–1793.

Lenton, T. M., Myerscough, R. J., Marsh, R., Livina, V. N., Price, A. R., Cox, S. J., & Team, G. (2009). Using GENIE to study a tipping point in the climate system. *Philosophical Transactions of the Royal Society A: Mathematical, Physical and Engineering Sciences, 367*(1890), 871–884.

Ligtvoet, W., & Knoop, J. (2014). *Kleine kansen—grote gevolgen: Slachtoffers en maatschappelijke ontwrichting als focus voor waterveiligheidsbeleid.* https://www.pbl.nl/publicaties/kleine-kansen-grote-gevolgen

Limaheluw, J., Schets, C., Vermeulen-Henstra, L., De Roda Husman, A. (2023). Klimaatverandering en wateroverdraagbare infectieziekten in Nederland. *Nederlands Tijdschrift voor Medische Microbiologie*, 31(2). https://www.nvmm.nl/media/5259/21714-054.pdf

LNV. (2020). Actieprogramma klimaatadaptatie landbouw. *Ministerie van Landbouw, Natuur en Voedselkwaliteit.* https://open.overheid.nl/documenten/ronl-35db3cf1-e9e6-4663-8963-4844da9ce1d1/pdf

Mulder, A. C., Pijnacker, R., de Man, H., van de Kassteele, J., van Pelt, W., Mughini-Gras, L., & Franz, E. (2019). "Sickenin' in the rain"—increased risk of gastrointestinal and respiratory infections after urban pluvial flooding in a population-based cross-sectional study in the Netherlands. *BMC Infectious Diseases, 19*(1), 377. https://doi.org/10.1186/s12879-019-3984-5

Nationaal Deltaprogramma. (2024a). *Advies Signaalgroep Deltaprogramma 2024.* https://www.deltaprogramma.nl/documenten/publicaties/2024/11/29/advies-signaalgroep-deltaprogramma-2024

Nationaal Deltaprogramma. (2024b). *Naar een nieuwe balans in de leefomgeving: Ruimte voor leven met water.* https://dp2025.deltaprogramma.nl

NIPV. (2019). *Maatschappelijk herstel na grootschalige overstromingen. Nationaal Instituut Publieke Veiligheid.* https://nipv.nl/wp-content/uploads/2022/02/20191017-IFV-Wave2020-Maatschappelijk-herstel-na-grootschalige- overstromingen.pdf /

NIPV. (2023). Natuurbrandsignaal '23. Nationaal Instituut Publieke Veiligheid. https://wp-content/uploads/2023/04/20230123-NIPV-Natuurbrandsignaal-23.pdf

O'Brien, K. (2009). Do values subjectively define the limits to climate change adaptation. *Adapting to climate change: Thresholds, values, governance*, 164, 180.

Overheid Wettenbank. (2019). *Klimaatwet.* https://wetten.overheid.nl/BWBR0042394/2020-01-01

PBL. (2015). Aanpassen aan klimaatverandering—Kwetsbaarheden zien, kansen grijpen (1454). *Planbureau voor de Leefomgeving.* https://www.pbl.nl/publicaties/aanpassen-aan-klimaatverandering-kwetsbaarheden-zien-kansen-grijpen

PBL. (2024). Klimaatrisico's in Nederland: De huidige stand van zaken (5359). *Planbureau voor de Leefomgeving.* https://www.pbl.nl/system/files/document/2024-05/pbl-2024-klimaatrisicos-in-nederland-5359.pdf

Rijksoverheid. (2023). *Beleidstafel wateroverlast en hoogwater.* https://open.overheid.nl/documenten/dpc-327e4ad8c480deec85c2c5b625fa723b8554d0e5/pdf

Rijksoverheid. (2024). *Nog genoeg ruimte om te bouwen, maar kies verstandige locaties.* https://www.rijksoverheid.nl/actueel/nieuws/2024/04/09/nog-genoeg-ruimte-om-te-bouwen-maar-kies-verstandige-locatiesnipv

SCP. (2025). Klimaat en Samenleving—Burgerperspectieven. *Sociaal en Cultureel Planbureau.* https://www.scp.nl/publicaties/publicaties/2025/04/17/klimaat-en-samenleving

Steg, L. (2023). Psychology of Climate Change. *Annual Review of Psychology*, 74, 391–421. https://doi.org/10.1146/annurev-psych-032720-042905

Storymaps. (n.d.). *Ruimtelijk afwegingskader klimaatadaptieve gebouwde omgeving.* https://storymaps.arcgis.com/stories/e14fe0614cc1440496b90ae03e7a2ce0

Van der Steen, M., Scherpenisse, J., De Jong, I., Boonsta, F., & Arnouts, R. (2016). *De volgende stap. Sturen met maatschappelijke energie in het natuurdomein.* Nederlandse School voor Openbaar Bestuur (NSOB).

Van Duin, M. (1992). *Van rampen leren: een vergelijkend onderzoek naar de lessen uit spoorwegongevallen, hotelbranden en industriële ongelukken.* Haagse Drukkerij.

Van Thiel, L., & Mol, P. (2020). Waterbewustzijn van de Nederlandse bevolking. *In opdracht van Kantar en HWBP.* https://klimaatadaptatienederland.nl/@237838/waterbewustzijn-van-de-nederlandse-bevolking/

Van Valkengoed, A. M., & Steg, L. (2019). Meta-analyses of factors motivating climate change adaptation behaviour. *Nature Climate Change, 9*(2), 158–163.

Van Valkengoed, A. M., Perlaviciute, G., & Steg, L. (2022). Relationships between climate change perceptions and climate adaptation actions: policy support, information seeking, and behaviour. *Climatic Change, 171*(1), 14. https://doi.org/10.1007/s10584-022-03338-7

Van Westen, R. M., & Dijkstra, H. A. (2023). Asymmetry of AMOC Hysteresis in a State-of-the-Art Global Climate Model. *Geophysical Research Letters, 50*(22), e2023GL106088.

Veiligheidsregio Limburg-Noord, & Veiligheidsregio Zuid-Limburg. (2020). *Rampbestrijdingsplan Hoogwater Maas Limburg 2020–2023.* https://lokaleregelgeving.overheid.nl/CVDR719417/1

VenW. (2009). Nationaal Waterplan 2009–2015. *Ministerie van Verkeer en Waterstaat.* https://zoek.officielebekendmakingen.nl/blg-47417.pdf

Verenigde Naties (VN). (2015). *Paris Agreement.* United Nations treaty collect, 1–27. https://unfccc.int/sites/default/files/english_paris_agreement.pdf

Verordening (EU). (2021). 2021/1119 van het Europees Parlement en de Raad van 30 juni 2021 tot vaststelling van een kader voor de verwezenlijking van klimaatneutraliteit, en tot wijziging van Verordening (EG) nr. 401/2009 en Verordening (EU) 2018/1999 ("Europese klimaatwet"), 1–17. https://climate.ec.europa.eu/eu-action/european-climate-law_nl

Witte, J. P. M., De Louw, P. G., van Ek, R., Bartholomeus, R. P., Van den Eertwegh, G., Gilissen, H. K., Beugelink, G., Ruijtenberg, R., & Van der Kooij, W. (2020). Aanpak droogte vraagt transitie waterbeheer. *Water Governance, 3*, 120–131.

WKR. (2023). *Met iedereen de transities in Richtinggevende keuzes voor een klimaatneutraal en klimaatbestendig Nederland.* https://www.wkr.nl/documenten/rapporten/2023/12/15/adviesrapport-met-iedereen-de-transities-in

WRR. (2008). Onzekere veiligheid (rapport nr.82). *Wetenschappelijke Raad voor het Regeringsbeleid.* https://www.wrr.nl/publicaties/rapporten/2008/10/01/onzekere-veiligheid

WRR. (2023). Goede zaken: Naar een grotere maatschappelijke bijdrage van ondernemingen (rapport nr.107). *Wetenschappelijke Raad voor het Regeringsbeleid.* https://www.wrr.nl/publicaties/rapporten/2023/09/14/goede-zaken

Chapter 3
How People Deal with the Uncertainty of Climate Threats

Contents

Key Points

- People respond differently to the uncertainty of climate threats, due mainly to variations in both threat appraisal and coping appraisal.
- Sociological and socio-psychological factors, such as group identities and perceived ability to implement measures, shape threat appraisal and coping appraisal.
- Climate events that have already occurred can influence threat appraisal and coping appraisal, both positively (self-confidence) and negatively (tunnel vision, trauma, resentment).

3.1 Engaging with Inherent Uncertainty

Adaptation to a changing climate raises questions about how citizens and communities deal with inherent uncertainty. This uncertainty has multiple aspects. First, it is uncertain what they will experience from climate change and when. There is also uncertainty about what others are doing to address threats and what can be expected

R. El-Dardiry et al., *People and Climate*, Research for Policy,
https://doi.org/10.1007/978-3-032-25831-1_3

from public and private parties. Given all this, people may also feel uncertain about what they themselves can or should do.

People and societies deal with uncertainty in very different ways. Some people are inclined to take precautionary measures at the earliest sign of threat, while others are more likely to accept risks as a part of life. Minimising the likelihood or extent of damage is not automatically the primary consideration. 'The facts' rarely speak for themselves.

A striking illustration of this can be found in the Lob van Gennep ('Gennep plain'), an area in the north of the Dutch province of Limburg, near the border with Germany. The area is enclosed on one side by a push moraine formed during the penultimate Ice Age and on the other by two rivers, the Meuse and the Niers. Around 7,000 people live scattered across five villages. Its location makes the area susceptible to flooding. Around Christmas 1993, high water in both rivers led to dangerous situations, both in the Lob van Gennep and in the neighbouring town of Gennep itself. A dyke along the Niers broke, a state of emergency was declared, and part of the population had to be evacuated. In 1995, water levels in the Meuse and the Niers were again very high. Gennep remained dry this time, but the residents of the nearby village of Middelaar were strongly advised to leave their village.[1]

By 2050, everyone living behind a primary flood defence in the Netherlands should have a basic level of protection, partly in anticipation of climate change.[2] The probability of death as a result of flooding must not exceed an average of one every 100,000 years. This level of protection has been translated into flood protection standards for each dyke section.[3] The current dykes in the Lob van Gennep do not meet these protection standards. For this reason, the authorities put forward three proposals in 2020 to improve flood safety in the area.[4] Option 1 involved a standard dyke raising, so that the minimum protection standard would be met. Option 2 entailed a slightly greater raising, except at three locations, referred to as 'thresholds'. These would make it predictable where water would enter the area in the event of extremely high water levels. Option 3 provided for additional raising and reinforcement of the dykes and the installation of a 'water retention inlet facility' that could be opened during extremely high water levels in the Meuse. In such a scenario, the five villages in the Lob van Gennep would be flooded in an extreme case, but larger downstream towns and cities, such as Den Bosch, would be spared. The Lob van Gennep would then serve as an overflow area.

Option 3 theoretically offered the residents of the Lob van Gennep the highest level of protection. The probability of flooding would be reduced to once every 3,000 years instead of once every 300 years. Nevertheless, the plan led to public

---

[1] De Gelderlander (1995).

[2] National Deltaprogramma (n.d.).

[3] Depending on the number of inhabitants and the potential damage that could result from a flood, the flood protection standard may vary for each dyke section. The protection standard for the dykes of the Lob van Gennep has been set at once every 300 years. The protection standard for dyke sections increases as the potential for large numbers of casualties or significant economic damage from flooding rises.

[4] Lob van Gennep Steering Committee (2020).

commotion, as residents focused on the harmful impact of a worst-case scenario in option 3: the opening of what became popularly known as 'the sluice'. Citizens organised themselves, including through the foundation 'No to the deluge'. One entrepreneur predicted that all businesses would go bankrupt at once if the sluice were ever opened.[5] The potential consequences of this felt unacceptable and unjust to many residents. Why should they be put at risk to guarantee dry feet for city dwellers living downstream? Ultimately, option 3 was not chosen: the government decided to reinforce the regular dykes in the area instead. Although this would produce a smaller flood safety gain than option 3 would, it received broader public support.[6] As a result, community tensions in the Lob van Gennep appear to have somewhat subsided.[7]

The Lob van Gennep case is not an isolated one. In this chapter, we demonstrate that socio-psychological factors play a major role in dealing with uncertainty in times of climate change. We first introduce a conceptual framework (Sect. 3.2). We then discuss how people perceive climate threats and what they think of the implementation of adaptation measures. What citizens expect from the government influences their own willingness to act (Sect. 3.3). We then analyse how specific experiences with climate events affect adaptation behaviour (Sect. 3.4). Illustration 3.1 demonstrates how people deal with uncertain climate threats, such as rising river levels.

3.2 A Conceptual Framework: Threat Appraisal and Coping Appraisal

Behavioural scientists have studied how people anticipate uncertain situations and, more specifically, potential hazards. These studies have resulted in widely accepted theories such as the 'protection motivation' theory and its later variants.[8] Two factors recur in these various socio-psychological perspectives (Fig. 3.1).[9]

The first important factor is *threat appraisal*: how are hazards perceived? The course of the coronavirus pandemic in the Netherlands illustrates how risk perception within a society can change and diverge.[10] Some saw the virus as merely 'a bad cold', while others were very concerned for their own health or that of loved ones and wanted to avoid infection at all costs. If people conclude that a particular hazard will not affect them because it will blow over, or because they think it will have relatively little impact on them, they are less motivated to adjust their lifestyle or to take targeted preparatory measures.

---

[5] Eimers (2020); Castricum (2020).

[6] Lob van Gennep (2022).

[7] Ariaans (2024a, 2024b).

[8] Rogers (1975); Witte (1992).

[9] Morss et al. (2024).

[10] Dryhurst et al. (2022); Van der Laan et al. (2022)

Illustration 3.1 Dealing with uncertainty

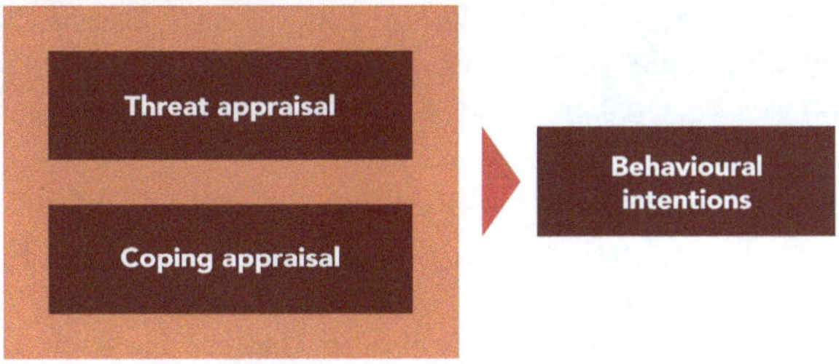

Fig. 3.1 Protection motivation theory[11]

Cognitive distortions of reality (*biases*) influence how people perceive uncertain situations.[12] For example, people tend to think from the perspective of the current situation. This *normalcy bias* can lead people to downplay or even completely overlook the pace and scope of impending changes and dangers.[13] People also look at what those around them say and do. Fear among close contacts or exposure to intense news reports, for instance, can heighten the perception of threat.[14] People also unconsciously attach more value to information and judgments from others that confirm their own intuitive interpretations (*confirmation bias*).[15]

Secondly, perceptions of the ability to mitigate danger—*coping appraisal*—also play a role. This concerns questions such as: Can the risk be reduced through targeted actions (outcome efficacy)? And if so, what can I do myself (self-efficacy)? What would it cost me? Here, too, people can develop very different perspectives. If confidence in one's own abilities is low, individuals will be less inclined to take action themselves to mitigate risks.[16] People may also be reluctant to act themselves if they initially rely on the government for protection.[17] Moreover, people tend to perceive it to be difficult to make long-term choices if these entail costs or inconvenience in the short term.[18]

In making both types of assessments, people are sensitive to what they hear from others in their environment or observe others doing. The social norms and practices of particular groups or communities can either facilitate or hinder the acceptance of measures to mitigate risk. Seeing neighbours, family, and friends take measures—or refrain from doing so—provides reference points for one's own behaviour. This was demonstrated in the first weeks of the coronavirus pandemic, when wearing face masks quickly became a new social norm in the Netherlands. The Dutch population largely suspended customary greeting etiquette such as handshakes and kisses on the cheek. In later stages of the pandemic, risk perception and belief in the effectiveness

---

[11] Cf. Marikyan and Papagiannidis (2023).

[12] WRR (2014).

[13] WRR (2022).

[14] Kasperson et al. (1988); Pidgeon and Henwood (2010).

[15] Nickerson (1998).

[16] WRR (2023).

[17] Oakley et al. (2020).

[18] Thaler and Shefrin (1981).

of prescribed measures diverged more, partly as a result of differences in the extent to which people trusted official communications. As a consequence, compliance with restrictive measures and willingness to be vaccinated markedly declined in some sections of the population.[19]

It is important for policymakers to understand this socio-psychological dynamic surrounding perceptions of and responses to risks. The old sociological dictum 'If men define situations as real, they are real in their consequences' is certainly applicable here.[20] People will not take adaptive action in response to dangers or opportunities they do not perceive or cannot interpret, nor will they act if they believe their actions will be of little use or if they are unsure what is expected of them. Moreover, they will be unlikely to support the policies and communications of institutions in which they have little trust.

3.3 Sociological and Socio-Psychological Factors in Adaptation Behaviour

Climate change is a fact,[21] but the how, where, and when of specific climate events are inherently uncertain. The extent to which people consider various manifestations of climate change to be significant for their own lives and take action to adapt to it is determined by the ways in which they deal with these uncertainties. In this sense, climate risks cannot simply be interpreted as probability x impact scores that supposedly represent developments in the physical world. They are also, to a significant extent, social constructs that are defined by the ways people think and talk about them. As Paul Slovic, one of the pioneers of behavioural decision-making science, once aptly put it: 'Risk does not exist independent of our minds and our culture.'[22]

Experts are gaining increasing insight into the 'objective' climate risks that people face. For the Netherlands, for example, climate vulnerability maps have been developed that indicate the magnitude of climate risks in specific neighbourhoods.[23] It has also become clear that certain population groups are more vulnerable.[24] For instance, older adults living in social isolation with multiple chronic conditions and limited mobility, newborn babies, and people living in poverty are at the greatest risk during heatwaves.[25] We will return to this in detail in Chapters 4 and 5.

---

[19] Cf. Gijsberts et al. (2022); Snel et al. (2022).

[20] Thomas (1928).

[21] IPCC (2023).

[22] Slovic (1992); Thomella et al. (2015).

[23] Climate Impact Atlas (n.d.)

[24] Kind et al. (2020).

[25] Klinenberg (2002); UNICEF (2024).

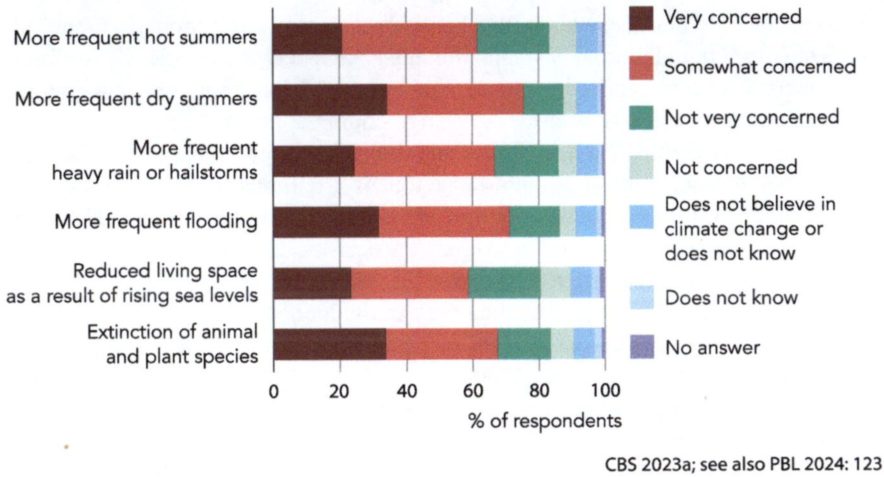

Fig. 3.2 Concerns about climate risks among Dutch citizens

However, whether citizens arrive at similar assessments and subsequently adjust their behaviour accordingly is questionable.[26] Citizens' perceptions of climate risk are selective. In the previous chapter, we already referred to the 2023 Statistics Netherlands study, which shows that a large majority of Dutch people have at least some concerns about climate threats such as rising temperatures (Fig. 3.2) and their implications for water safety. At the same time, relatively few Dutchmen are concerned about risks related to heat stress, which is a relatively unfamiliar phenomenon in the Netherlands and thus less easily imagined and understood phenomenon.[27]

Furthermore, the step from perception to behaviour is not self-evident.[28] In this context, Bechtoldt et al. refer to an *adaptation puzzle*: why do people fail to take adaptation measures that are clearly rational for them from both an economic and safety perspective?[29] This is also referred to as a *climate dilemma*: while the vast majority of Dutch people are concerned about climate change and its manifestations, increasingly hot summers, rising sea levels, and declining biodiversity, their mitigation and adaptation behaviour lags behind.[30]

There are significant variations in how people assess climate risks and in their motivation to adjust their behaviour accordingly. These differences manifest themselves at the international level, for example between residents of different EU

---

[26] Cf. Meyer and Kunreuther (2017); Nationaal Deltaprogramma (2024).

[27] CBS (2023); Frederik (2024).

[28] Nationaal Deltaprogramma (2024).

[29] Bechtoldt et al. (2020).

[30] Renes (2021); SCP (2025).

Member States,[31] but also nationally: between different generations,[32] urban and rural residents[33] mainland and island inhabitants,[34] and between people with different political preferences[35] and educational attainment.[36] Moreover, this concerns not only differences in threat appraisal, but also in coping appraisal.[37]

The use of metaphors such as puzzle and dilemma suggests that the relationship between perception, motivation, and behaviour is a complex one. People may not take action to mitigate the risks they run for a wide variety of reasons. They may be unaware of the risks they face, downplay these risks, consider it too expensive or complicated to take action themselves, trust that others or authorities will act, and so on. Moreover, there is more at play than their individual cognition and motivation: a big body of knowledge into perceptions of climate risks and people's willingness to take adaptive action or to accept adaptation measures taken by the authorities show that multiple sociological and socio-psychological factors are at play.[38] Four such factors are particularly important.

The first is *group memberships and identities*. The groups to which people belong (or to which they would like to belong) partly determine how they view the world. Group standards and values also influence what people consider worth preserving and protecting, what they recognise as dangers, and what they are willing to give up or do to mitigate perceived risks. People's risk perceptions can be influenced by various social identities, such as ethnicity, income, nationality, political beliefs, and gender.[39]

Second, the extent to which people believe that their *own behaviour and that of others in the community can be effective* in mitigating the risks they collectively face plays a role.[40] Van Valkengoed et al. studied the role of perceptions of climate risks and the expectations that one's own adaptation behaviour can contribute to the desired outcome. They argue that both mediate the relationship between the assumption that climate change is a real phenomenon and actually engaging in climate adaptation behaviour. A related study goes a step further and suggests that it is not so much the perceived effectiveness of individual behaviour that matters, but rather that of collective behaviour. In this comparative study between the United States and the Netherlands, the researchers show that the more people believe in the adaptation

[31] Heimann and Mallick (2016); Bouman et al. (2020).

[32] Poortinga et al. (2023).

[33] Tenbrink and Willcock (2023).

[34] Tezar and Setiadi (2023); Abu Samah and Shaffril (2020).

[35] Schwaller et al. (2020).

[36] Educational attainment has been described by some researchers as 'worldwide the single strongest predictor of climate change awareness'; see Lee et al. (2015).

[37] WRR (2023); Van Valkengoed and Steg (2019a, 2019b); Bechtoldt et al. (2020).

[38] Van der Linden (2015, 2017); Van Valkengoed and Steg (2019a, 2019b).

[39] Mackay et al. (2021).

[40] Van Valkengoed et al. (2024); Nasi et al. (2024).

capacity of the community as a whole, the more likely they are to take adaptive action themselves.[41]

A third important factor is *the degree of trust people have in the problem-solving capacity of institutions.* People derive institutional trust in climate policy based on their judgments about the integrity and competence of experts and civil-society organisations active in this field.[42] This also applies to the actions of government agencies. People have more trust in official measures when they believe the authorities pursue the same values as they do, and in measures that have been developed in processes that they consider to have been fair.[43] High trust in the climate policy of institutions can have various effects on people's own behaviour. They may come to rely on it to such an extent that their threat appraisal diminishes, along with their motivation to take adaptive action in their own lives. Conversely, they may become more responsive when the authorities call on them to take action.

Fourth, people develop *rationales and cultural practices* that enable them to accept living in risky areas, so that they do not have to give up their sources of income or social identity. In effect, people weigh the risks of climate events against the benefits of living in that area. They find support for this among neighbours, colleagues, and fellow believers.[44] Consider the folkloric or spiritual rituals through which agricultural communities are able to embrace unpredictable weather conditions and variable crop yields. International comparative research in 46 non-industrial communities shows that residents of countries in a dry climate were more likely to believe that divine powers could provide deliverance.[45] This also occurs in Western, secularised countries. For example, during a prolonged drought in Australia that placed a heavy burden on rural residents and led to a sharp increase in suicides among farmers, churches filled with people praying for rain.[46]

3.4 Socio-Cultural Impacts of Climate Events

In this section, we shift perspective and examine how past climate events continue to influence perceptions of climate threats and coping appraisal. Disasters can cast long shadows over communities. What people have previously experienced in terms of natural disasters, what they tell each other and others about these events, and how these are recorded in collective memory, all shape their ideas about danger and their willingness to invest in their own safety. Research into trauma and the political and policy processing of crises confirms this picture.[47] Communities can develop active

[41] Nasi et al. (2024).

[42] Cologna and Siegrist (2020).

[43] Kitt et al. (2021); Liu et al. (2020); Palomo-Vélez et al. (2024).

[44] Cannon (2015).

[45] Ember et al. (2021).

[46] ACBC (2019).

[47] Gerson (2024); Stark (2018).

cultures of remembrance and preparedness practices, especially when disasters occur repeatedly (see also Box 3.1).[48]

Box 3.1 Path-dependent adaptation practices on the German Baltic coast
A series of studies into climate adaptation processes along the Baltic coast of Germany highlights a striking contrast between two communities, both of which had to recover in the aftermath of a devastating storm surge in 1872. Both communities now face the effects of climate change, such as rising sea and groundwater levels and an increased risk of flooding. However, the local residents have developed very different cultural identities. These identities influence their attitudes towards their natural environment and the sea they live near, the risks that climate change poses for them, and the importance of both material and immaterial values in their assessment of possible adaptation measures.

On the low-lying and partly reclaimed island of Ummanz (20 km^2; approximately 250 inhabitants), which is connected by a bridge to the much larger island of Rügen, residents have been exposed to natural hazards for as long as anyone can remember. The memory of the 1872 disaster is deeply rooted in the collective memory and mentality of the islanders. Stories about it are passed down from generation to generation, and various artefacts such as memorials serve as reminders. The 1872 storm surge, the construction of a protective sea wall in response, the subsequent development of agriculture, and—since German reunification—the rise of boutique ecotourism form the core of a cultural myth that a protected and comfortable life in this remote place of natural beauty will remain possible in the future.

The current sea wall is in a poor state of repair and inadequate in light of the expected effects of climate change in the region. However, the coastal protection plan proposed by the regional government is seen by the islanders as a threat to their hard-won identity as a rural community living in harmony with nature. The municipality cannot afford the residents' preferred option of reconstructing the old sea wall. Instead, the regional government wants to build a new dyke along the axis of the island. As a result, some parts of the island will be returned to the sea, while the inhabited areas will be better protected. The residents see this as a direct attack on their way of life. This has resulted in a stalemate that hinders the implementation of measures to mitigate risk.

In Timmendorfer Strand, located about 150 kms further west, there is a very different attitude. The 1872 storm surge there led to a different socio-economic development. A spa and seaside resort has made tourism the main source of income. This has also strongly influenced the local identity and mentality.

[48] Schenk (2015).

Box 3.1 (continued)

The prevailing values of today's residents are material. They see the Baltic Sea, the beach, and the coast as the economic lifeline of their community and take the threat of flooding extremely seriously, despite perceiving the likelihood as relatively low. The community's driving motivation is the protection of investments and the associated future prospects of the residents. Moreover, unlike in Ummanz, the municipality has given residents ample opportunity to influence decision-making on local adaptation measures. Together, these factors have greatly contributed to the development and acceptance of a local adaptation strategy in which the values of coastal protection and the promotion of tourism are intertwined.[49]

Clearly, communities can learn different lessons from the same climate events. Past experiences can influence the motivation to shape protection, depending on how people and authorities deal with periods of acute danger and loss. Based on studies of the long-term effects of disasters, we distinguish four patterns: self-confidence, tunnel vision, trauma, and resentment.[50]

3.4.1 Self-Confidence

In the pattern characterised by self-confidence, crisis experiences lead to feelings of solidarity: 'we' have been tested and strengthened by the disaster, and have emerged stronger by standing shoulder to shoulder. The dominant lesson drawn from this is that 'our' resilience to adapt to climate events is considerable. The collective perceives a strong capacity to act.

An example is how centuries of living in the Dutch delta have been embedded in a narrative of a 'battle against the water' that 'we' have learned to win. Some of the flood disasters that have occurred over the centuries are ingrained in the collective memory of the Dutch: the St. Elizabeth's flood in 1421, the storm surge in 1825, and the North Sea flood of 1953. These serve as reminders of the danger: the sea not only gives, but also takes away from time to time. However, the self-confidence narrative emphasises the control the Dutch have managed to gain over the water, for example in the image of the Dutch lion defeating the water wolf. The modern achievements that are consistently cited include the damming of the former Zuiderzee (now Lake IJssel), the Delta Works,[51] the extensive system of river dykes and pumping stations, and now also the adaptive water management embodied in the success of the government

[49] Martinez et al. (2020); Hallin et al. (2021).

[50] Arcaya et al. (2020); Grisham et al. (2023); Hettige (2023); Peters (2021).

[51] The Delta Works are a vast Dutch defence system of dams, sluices, and storm surge barriers, constructed to protect the Netherlands from coastal flooding.

programme Room for the River (see Box 3.2).[52] The institutions that underpin these efforts—the water authorities, the Directorate-General for Public Works and Water Management, and the independent Delta Programme Commissioner—are widely respected.[53]

Box 3.2 The government programmes Room for the River and Meuse Works

In the winters of 1993 and 1995, the rivers Meuse and Rhine were faced with enormous volumes of water heading towards the sea. The dykes along the rivers just about held, but there was significant flooding in the areas outside the dykes. This near-disaster was the impetus for the Room for the River programme.

In centuries past, rivers were shortened by embanking bends and meanders and they were narrowed to reclaim more land along the rivers and to speed up the discharge of water. Flood safety was achieved by making the dykes higher and stronger. The Room for the River and Meuse Works programmes marked a break with this trend. What was special about these programmes was the dual objective—ensuring flood safety and improving spatial quality—and the role of residents. Decision-making on the measures for Room for the River took place within the framework of a key planning decision, in which public participation was an integral part.

The measures themselves included relocating dykes, lowering stone breakwaters in rivers, removing obstacles in the floodplains, deepening the floodplains, and creating side channels. This created more space for water to be discharged safely and gave nature ample opportunity to develop along the river branches. The measures for the Meuse Works programme include widening and deepening the river, lowering floodplains, and constructing high-water channels and retention areas. These measures also create space for nature development.

The programmes started in 2005 on the river Meuse and in 2006 along the branches of the river Rhine. In total, interventions have been carried out at 34 locations along the Rhine branches. With the completion of the Reevediep in 2019, the Room for the River programme for the Rhine branches was concluded, while the Meuse Works programme continues for the time being.[54]

The narrative of control and mastery continues to influence the current approach to the consequences of climate change. The belief that nature can be controlled has been a robust component of Dutch climate adaptation policy: even in times of rising sea levels and changing weather patterns, the Netherlands can 'win' against nature through a combination of technical ingenuity and adjustments to the physical living environment.

---

52 Van Buuren (2019).

53 Van den Brink (2021).

54 Rijkswaterstaat (n.d.)

Such historically grounded strong collective self-confidence is a double-edged sword. It can certainly drive motivate proactive adaptation. But it can also produce a *vulnerability paradox*. In communities that feel protected by high levels of infrastructural preventive measures, risk awareness and the motivation to engage actively in mitigation and preparedness may be relatively underdeveloped.[55] In such communities, so much of the once-feared danger has been brought under control that many people have come to expect comprehensive and flawless protection. As a result, they lose sight of the residual and newly emergent risks they actually do face. Such neglect of residual risks makes citizens vulnerable when disaster scenarios that are no longer feared—because they have widely become considered impossible—do in fact materialize.[56] In line with this, Dutch linguist and philosopher Lotte Jensen observes that '[i]f you assume that there is a technological fix for every problem, the need to examine human behaviour critically disappears.'[57]

As an example of this, consider the protective measures implemented after 1995 around the villages of Itteren and Borgharen in the southern part of the province of Limburg. While these measures significantly reduced flood risks, they also created the potential for a 'bathtub effect' in the event of a dyke breach or extreme water levels. This refers to the rapid and high inundation of a low-lying area surrounded on all sides by dykes, thereby increasing risks to people and animals, significantly reducing available evacuation time, and causing greater damage to buildings.[58] It is highly questionable whether the local population sufficiently recognises the problem that this 'solution' poses.[59]

3.4.2 Tunnel Vision

Communities with experience of disasters are better prepared for 'the devil they know' than for unknown but nonetheless real threats. In communities that have primarily dealt with a particular type of threat due to their location and history, preparedness for that specific event is well developed. Consider high water in river deltas and wildfires in Australia and southern European countries. Hazards that are not kept relevant by personal experience or collective memory resonate far less in disaster planning. This bias in the 'risk catalogue' comes at the expense of preparedness for relatively unfamiliar risks that these communities face.[60]

Most Dutch adults learned in geography class that the Netherlands has a temperate maritime climate, or something similar. This already puts Dutch citizens at a disadvantage when trying to grasp the significance of the fact that, due to climate change,

[55] For the Netherlands, see Terpstra (2011); Netherlands Court of Audit (2023).

[56] Rosenthal (1988); De Graaf (2022).

[57] Jensen (2020: 13).

[58] Kasmalkar et al. (2024).

[59] Universitat Autònoma de Barcelona (n.d.)

[60] Diakakis et al. (2021); Etkin et al. (2018).

even temperate climate zones can experience much more severe storms. The natural disaster scenario that most Dutch people are alert to is that of 'too much water in the wrong place.' This remains highly relevant in times of climate change, but at the same time, at least two of the four dominant climate threats identified by the Royal Netherlands Meteorological Institute—more heatwaves and prolonged droughts—do not receive equal attention. Box 3.3 illustrates how such blind spots can cause problems.

Box 3.3 Unprepared for the silent killer called heat[61]

Between 4 and 18 August 2003, unusually high temperatures combined with socio-economic vulnerability led to nearly 15,000 excess deaths in France. Older people were the most vulnerable, but excess mortality was also observed in other age groups. For men aged 35 to 44, excess mortality exceeded 20%. The main causes of death were dehydration, hyperthermia, and heatstroke.

Most of the excess deaths occurred in urban centres, such as the Paris region. Research has shown strong correlations with the living conditions of the victims, particularly poverty, isolation, and poor health.[62] Of the victims, 92% lived alone and 41% lived in a one-room apartment, with an area smaller than 10 m^2 in 12% of cases. Police and fire services, often the first to arrive after a report, consistently described the stifling temperatures in the rooms where the bodies were found.

Other European countries also experienced historic temperature peaks in August 2003, ranging from 38.5 °C (in England and Wales) to 47 °C (in Portugal). England and Wales saw an excess mortality of 16% between 4 and 13 August. Portugal, on the other hand, experienced much higher temperatures but a comparatively low increase in mortality (26%) throughout the entire month of August compared to the same period in the years 1997–2001. This was not a coincidence, but the result of adaptation policy. For example, in 1999, the city of Lisbon established a preventive health service in anticipation of future heatwaves, which in 2003 was actively engaged in early warning and the mobilisation of protective measures and social support for at-risk groups. In France, there was virtually no preparation of any kind. As a result, heat became a predictable but neglected silent killer.

The 2003 heatwave in France served as a wake-up call that heat could no longer be ignored as a climate phenomenon that regularly affects the country and, as a consequence of climate change, will do so with increasing frequency. The impact of subsequent heatwaves has proven to be substantial. One study has estimated the economic impact of selected health effects of heatwaves in France during the period 2015–2019 at €25.5 billion, mainly in the areas of

[61] This box text is based on the publication by Poumadere et al. (2005).

[62] Keller (2015); Adélaïde et al. (2022).

Box 3.3 (continued)
mortality (€23.2 billion), reduced active days (€2.3 billion), and morbidity (€0.031 billion).[63] Heatwaves thus represent a significant economic burden for the French health care system and a considerable health risk for segments of the population.

3.4.3 Trauma

A trauma pattern can emerge when people's disaster experiences make them painfully aware of the magnitude of the dangers they face and of their limited capacity to protect themselves.[64] In this pattern, feelings of loss, damage, and lack of control are predominant. The social fabric of affected communities can suffer severe damage from traumatising disaster experiences, reflected among other things in the erosion of trust. This can provoke or reinforce a fatalistic attitude toward hazards: let the future take care of itself. As a result, the capacity to act is undermined further. It may also prompt people to escape the chronically vulnerable situation, ranging from 'internal migration'—suppressing what has happened and what may still occur—to physical migration.[65] Research among those affected by flooding in the province of Limburg in 2021, for example, has shown that anxiety about a possible future flood is a significant predictor of the intention to move to a location without flood risk.[66]

Elements of the trauma pattern can be found in the long-term community impacts of hurricane Irma, which struck the Caribbean islands of Sint Maarten, Saba, and Sint Eustatius—part of the Kingdom of the Netherlands—on 6 September 2017. Fully 90% of the buildings and virtually all infrastructure on Sint Maarten were damaged.[67] The island's residents generally demonstrated resilience in the immediate response phase, and there was considerable solidarity in providing assistance to those who had lost their homes and possessions. Emergency aid from the Netherlands was relatively swift, although tensions arose over time between the Dutch government and the

[63] Adélaïde et al. (2022).

[64] The social psychologist Hirschberger (2018: 1) offers a striking analysis of collective trauma: 'The term collective trauma refers to the psychological reactions to a traumatic event that affect an entire society; it does not merely reflect an historical fact, the recollection of a terrible event that happened to a group of people. (…) Collective memory of trauma is different from individual memory because collective memory persists beyond the lives of the direct survivors of the events, and is remembered by group members that may be far removed from the traumatic events in time and space. These subsequent generations of trauma survivors, that never witnessed the actual events, may remember the events differently than the direct survivors, and then the construction of these past events may take different shape and form from generation to generation.'

[65] Saul (2022).

[66] Duijndam et al. (2023).

[67] RVO (2025).

government of Sint Maarten—partly because the Dutch government imposed conditions regarding good governance.[68] Six to eight weeks after hurricane Irma, however, more than 30% of Sint Maarten's residents were suffering from post-traumatic stress symptoms, despite their initial resilience. Many decided to leave the island and seek refuge elsewhere: a year and a half after the disaster, the number of islanders who had migrated to the three largest cities in the European part of the Kingdom of the Netherlands had more than doubled compared to the period before the hurricane.[69]

3.4.4 *Resentment*

The final way in which climate events can leave their mark on the culture of a community is characterised by feelings of resentment. Criticism of the actions of authorities and disappointment over the lack of solidarity within society underlie these feelings. The prevailing narrative is that 'we' have been abandoned by everyone, or that 'they' (the authorities and other groups) apparently considered protecting other interests more important than addressing 'our' needs. The dominant sentiment is that authorities (and sometimes fellow citizens) cannot be trusted when it matters. The authorities are then widely blamed for the damage and suffering resulting from a new disaster. The aftermath of the climate disaster in the Spanish region of Valencia in 2024 provided a dramatic example of this. During a visit to the affected area, the king of Spain was pelted with mud and booed, and the regional governor became a widely despised figure, bereft of the legitimacy normally accorded to the holder of that office.[70]

Even if there is no existing pattern in society of fatalistic expectations, it is not uncommon for 'disasters after the disaster' to occur during the response and recovery phases of disasters. In the immediate response phase, everyone often puts their best foot forward and much can be arranged quickly, but the subsequent reconstruction and claims settlement processes are incomprehensibly and unacceptably bureaucratic for many: compartmentalised, formalised, and slow. This increases the losses suffered and exacerbates the sense of resentment among those affected. Dutch disaster psychiatrist Berthold Gersons refers to this as the disillusionment phase:

> The exhaustion and fatigue of the first weeks contribute to this. Victims and bystanders, even entire communities, risk losing their trust in authorities and government, sometimes for years. (…) [I]f victims are not properly supported and assisted after a disaster, a cascade of problems can arise.[71]

If such groups had already sensed that their concerns about increasing but unaddressed risks fell on deaf ears during the often years-long lead-up to the eventual

---

[68] Wijkhuis et al. (2018: 16); Inspectorate of Justice and Security / Inspectie JenV (2018: 58–59).

[69] BNNVARA (2017); Van der Sanden (2019).

[70] NOS (2024).

[71] Gersons (2024)

disaster, this can form a toxic cocktail. In cases such as the earthquakes in the northern Dutch province of Groningen, caused by gas extraction,[72] and the childcare benefits scandal,[73,74] we have seen the destructive effects of resentment not only on the well-being and resilience of affected groups, but also on social cohesion and the legitimacy of government more broadly.[75]

3.5 Conclusion

Climate change brings new uncertainties. In this chapter, we have seen how threat appraisal and coping appraisal influence how people deal with these uncertainties. The success of climate adaptation policy therefore partly depends on the extent to which these sociological and socio-psychological insights are taken into account.

Climate events and the subsequent administrative response can profoundly alter the mentality of communities and individuals. The Dutch have gained much self-confidence through their long and successful struggle against water. As the old Dutch saying goes: 'God created the earth—except for the Netherlands, for that was made by the Dutch themselves.' This self-confidence has brought the Netherlands much, but can also lead to overconfidence. In the Netherlands, attention almost automatically focuses on water-related hazards, while climate change also involves drought and heat. This attitude poses a risk to society. Belief in a certain degree of self-reliance and trust in fellow citizens and authorities can be severely shaken if climate events have disastrous outcomes and people are left disoriented. Traumatic experiences and feelings of resentment reduce the likelihood that people can, or want, to contribute to the common good.

The analysis in this chapter may give the impression that the sociological and socio-psychological factors discussed here are, in most cases, a hindrance to climate adaptation. However, this is not necessarily the case.

To understand this, we return to the Lob van Gennep. There, official plans led to increasing uncertainty among citizens and businesses. For many, the 'sluice' was a frightening prospect. However, the heated debates also led to renewed risk awareness.

---

[72] The parliamentary inquiry committee that investigated natural gas extraction in Groningen articulated the resentment felt by many residents in powerful terms. It referred to the 'systematically ignored interests' of inhabitants, 'unacceptably prolonged dangers', a 'wavering reinforcement strategy', and 'empty promises' in the recovery operation; see Parliamentary Committee of Inquiry on Natural Gas Extraction in Groningen / Parlementaire enquêtecommissie aardgaswinning Groningen. (2023).

[73] In the childcare benefits scandal, the Dutch Tax and Customs Administration wrongly accused and prosecuted thousands of parents between 2005 and 2019 for fraud with childcare benefits. This led to severe and, in many cases, catastrophic financial and personal problems for the affected families.

[74] Committee of Inquiry on Childcare Benefits / Parlementaire ondervragingscommissie kinderopvangtoeslag. (2020).

[75] Kanne and Driessen (2021).

Partly as a result, the village councils in the area began to explore what the community itself could do to enhance safety. Together, they drafted a memorandum proposing measures 'to increase awareness and self-reliance among residents.' They are also exploring 'collective forms of self-reliance.' Instead of each household putting up its own sand bags, for example, one could consider demountable walls that protect an entire street in case of emergency.[76]

In this way, the residents of the Lob van Gennep demonstrate that they are not powerless. A strong sense of mutual solidarity and a firm belief in their shared abilities allow communities to enhance their resilience. In such cases, and depending on the type of risk, the government may be better advised to take a supporting rather than a leading role. The next chapter will examine in greater depth the crucial role of social cohesion in dealing with climate events.

References

Abu Samah, A., & Shaffril, H. A. M. (2020). A comparative study between mainland and islander small-scale fishermen's adaptation towards climate change. *Environmental Science and Pollution Research, 27*, 11277–11289. https://doi.org/10.1007/s11356-019-07143-1

Adélaïde, L., Chanel, O., & Pascal, M. (2022). Health effects from heat waves in France: An economic evaluation. *The European Journal of Health Economics, 23*, 1–13.

Algemene Rekenkamer. (2023). *Voorbij de dijk. Keuzes in het waterveiligheidsbeleid*. https://www.rekenkamer.nl/publicaties/rapporten/2023/10/12/voorbij-de-dijk

Arcaya, M., Raker, E. J., & Waters, M. C. (2020). The social consequences of disasters: Individual and community change. *Annual Review of Sociology, 46*(1), 671–691.

Ariaans, J. (2024a, oktober 22). CDA Gennep wil opheldering: kan er nog gebouwd worden in Lob van Gennep? *AD*. https://www.ad.nl/gennep/cda-gennep-wil-opheldering-kan-er-nog-gebouwd-worden-in-lob-van-gennep~a1bfbb22/

Ariaans, J. (2024b, February 7). Inwoner Lob van Gennep moet huis zelf 'waterbestendig' maken. *De Gelderlander*.

Australian Catholic Bishops Conference (ACBC). (2019). *The Catholic bishops of Australia have announced a National Prayer Campaign for the gift of rain during November.* https://www.dow.org.au/wp-content/uploads/2019/10/Bulletin-Notices-OOB-26-27-October-2019-WEB.pdf

Bechtoldt, M. N., Götmann, A., Moslener, U., & Pauw, W. P. (2020). Addressing the climate change adaptation puzzle: A psychological science perspec- tive. *Climate Policy, 21*(2), 186–202. https://doi.org/10.1080/14693062.2020.1807897

BNNVARA. (2017). *ZEMBLA onthult: Dertig procent van bevolking Sint Maarten kampt met posttraumatische stressklachten na Irma*. https://www.bnnvara.nl/zembla/artikelen/dertig-procent-van-bevolking-sint-maarten-kampt-met-posttraumatische-stressklachten-na-irma

Bouman, T., Verschoor, M., Albers, C. J., Böhm, G., Fisher, S. D., Poortinga, W., Whitmarsh, L., & Steg, L. (2020). When worry about climate change leads to climate action: How values, worry and personal responsibility relate to various climate actions. *Global Environmental Change, 62*, 102061. https://doi.org/10.1016/j.gloenvcha.2020.102061

Cannon, T. (2015). Disasters, climate change and the significance of 'culture.' In H. Krüger, G. Bankoff, T. Cannon, B. Orlowski, & E. L. F. Schipper (Eds.), *Cultures and Disasters: Understanding Cultural Framings in Disaster Risk Reduction* (pp. 88–106). Routledge.

[76] Village Councils Working Group (2024).

Castricum, R. (presentator). (2020). *De Hofbar. De Lop van Gennep afvoerputje van Nederland?* https://www.youtube.com/watch?v=VE6zYTZZtZA

CBS. (2023). *Klimaatverandering en energietransitie: Opvattingen en gedrag van Nederlanders in 2023*. Centraal Bureau voor de Statistiek. https://longreads.cbs.nl/klimaatverandering-en-energietransitie-2023/

Cologna, V., & Siegrist, M. (2020). The role of trust for climate change mitigation and adaptation behaviour: A meta-analysis. *Journal of Environmental Psychology, 69*, Article 101428.

De Gelderlander. (1995, januari 30). Evacuatie op gang in Middelaar; Er mag geen auto het dorp meer in. *De Gelderlander*.

De Graaf, K. (2022). Naar een dynamische (Ed.) milieuvergunning. Bye-bye bestaande rechten? In *Milieurecht in transitie: Jubileumbundel 40 jaar VMR* (pp. 87–95). Boom Juridisch.

Diakakis, M., Skordoulis, M., & Savvidou, E. (2021). The relationships between public risk perceptions of climate change, environmental sensitivity and experience of extreme weather-related disasters: Evidence from Greece. *Water, 13*(20), 2842.

Dryhurst, S., Schneider, C. R., Kerr, J., Freeman, A. L., Recchia, G., Van Der Bles, A. M., Spiegelhalter, D., & Van Der Linden, S. (2022). Risk perceptions of COVID-19 around the world. In J. K. Wardman & R. Löfstedt (Eds.), *COVID-19* (pp. 162–174). Routledge.

Duijndam, S. J., Botzen, W. J. W., Endendijk, T., de Moel, H., Slager, K., & Aerts, J. C. J. H. (2023). A look into our future under climate change? Adaptation and migration intentions following extreme flooding in the Netherlands. *International Journal of Disaster Risk Reduction, 95*, 103840. https://doi.org/10.1016/j.ijdrr.2023.103840

Eimers, D. (2020, 22 April). Een tweede Groningen. *De Groene Amsterdammer*. https://www.groene.nl/artikel/een-tweede-groningen

Ember, C. R., Skoggard, I., Felzer, B., Pitek, E., & Jiang, M. (2021). Climate variability, drought, and the belief that high gods are associated with weather in nonindustrial societies. *Weather, Climate, and Society, 13*(2), 259–272.

Etkin, D. A., Mamuji, A. A., & Clarke, L. (2018). Disaster risk analysis part 1: The importance of including rare events. *Journal of Homeland Security and Emergency Management, 15*(2), 20170007.

Frederik, J. (2024). Hittedoden door klimaatverandering, wat weten we nu écht? *De Correspondent*. https://decorrespondent.nl/15521/hittedoden-door-klimaatverandering-wat-weten-we-nu-echt/7d8d0ef8-92ba-0e38-0d1c-3bc343bbda9e

Gerson, S. (2024). Affiliation vs. Alienation. *Psychoanalytic Dialogues, 34*(3), 280–287.

Gersons, B. (2024). Betoog: Voorkom op tijd de ramp na de ramp. *VNG Magazine, 3*. https://vng.nl/artikelen/betoog-voorkom-op-tijd-de-ramp-na-de-ramp

Gijsberts, M. et al. (2022). *Vaccinatiebereidheid en vertrouwen in instituties onder Nederlanders met en zonder migratieachtergrond*. https://stukroodvlees.nl/vaccinatiebereidheid-en-vertrouwen-in-instituties-onder-nederlanders-met-en-zonder-migratieachtergrond/

Grisham, E. L., Jones, N. M., Silver, R. C., & Holman, E. A. (2023). Do Past Events Sow Future Fears? Temporal Disintegration, Distress, and Fear of the Future Following Collective Trauma. *Clinical Psychological Science, 11*(6), 1064–1074.

Hallin, C., Martinez, G., Hofstede, J. L., Jensen, J., Baron, N., Heimann, T., Kroon, A., Arns, A., Almström, B., & Sørensen, P. (2021). A comparative study of the effects of the 1872 storm and coastal flood risk management in Denmark, Germany, and Sweden. *Water, 13*(12), 1697.

Heimann, T., & Mallick, B. (2016). Understanding Climate Adaptation Cultures in Global Context: Proposal for an Explanatory Framework. *Climate, 4*(4), 59. https://www.mdpi.com/2225-1154/4/4/59

Hettige, S. (2023). Introduction: Sociology of disasters. In A. Singh (Ed.), *International handbook of disaster research* (pp. 1529–1534). Springer.

Inspectorate of Justice and Security / Inspectie JenV. (2018). *Onderzoek naar het systeem van de rampenbestrijding op de BES-Eilanden*. Ministerie van Justitie en Veiligheid. 58–59. https://www.inspectie-jenv.nl/documenten/2018/06/27/reactie-minister-op-onderzoek-naar-het-systeem-van-de-rampenbestrijding-op-de-bes-eilanden

IPCC. (2014). *Summary for policymakers* (Climate Change 2014: Impacts, Adaptation and Vulnerability. Part A: Global and Sectoral Aspects. (Contribution of Working Group II to the Fifth Assessment Report of the Intergovernmental Panel on Climate Change). C. U. Press.

IPCC. (2022). *Climate Change 2022: Impacts, adaptation and vulnerability* (Working Group II Contribution to the Sixth Assessment Report of the Intergovernmental Panel on Climate Change). C. U. Press. https://www.ipcc.ch/report/ar6/wg2/

IPCC. (2023). Climate Change 2023. Synthesis Report. *Summary for Policymakers.* https://www.ipcc.ch/report/ar6/syr/downloads/report/IPCC_AR6_SYR_SPM.pdf

Jensen, L. (2020). De strijd tegen het water in woord en beeld. Verbindingen tussen heden en verleden. In L. Jensen, & A. Duiveman (Eds.), *Welke verhalen vertellen we? Narratieve strategieën rondom water- beheer en zeespiegelstijging* (pp. 6–13). Radboud Universiteit. https://repository.ubn.ru.nl/handle/2066/227293

Kanne, P., & Driessen, M. (2021). *Vertrouwen in overheid na drie grote kwesties* (2021/068). I&O Research.

Kasmalkar, I., Wagenaar, D., Bill-Weilandt, A., Choong, J., Manimaran, S., Lim, T. N., Rabonza, M., & Lallemant, D. (2024). Flow-tub model: A modified bathtub flood model with hydraulic connectivity and path-based attenuation. *MethodsX, 12*, Article 102524.

Kasperson, R. E., Renn, O., Slovic, P., Brown, H. S., Emel, J., Goble, R., Kasperson, J. X., & Ratick, S. (1988). The Social Amplification of Risk: A Conceptual Framework. *Risk Analysis, 8*(2), 177–187. https://doi.org/10.1111/j.1539-6924.1988.tb01168.x

Keller, R. C. (2015). *Fatal Isolation.* University of Chicago Press.

Kind, J., Koers, G., de Nijs, T., Koopmans, R., van Popering, J., Visser, V., & Damen, M. (2020). *Verkenning naar de wisselwerking tussen sociale veerkracht en klimaatadaptatie.* Deltares.

Kitt, S., Axsen, J., Long, Z., & Rhodes, E. (2021). The role of trust in citizen acceptance of climate policy: Comparing perceptions of government competence, integrity and value similarity. *Ecological Economics, 183*, Article 106958.

Klimaateffectatlas. (n.d.). *Klimaateffectatlas.* https://www.klimaateffectatlas.nl/nl/

Klinenberg, E. (2002). *Heat wave: A social autopsy of disaster in Chicago.* University of Chicago Press.

Lee, T. M., Markowitz, E. M., Howe, P. D., Ko, C.-Y., & Leiserowitz, A. A. (2015). Predictors of public climate change awareness and risk perception around the world. *Nature Climate Change, 5*(11), 1014–1020.

Liu, J., Zhao, M., & Wang, Y. (2020). Impacts of government subsidies and environmental regulations on green process innovation: A nonlinear approach, *Technology in Society*, 63, 101417, ISSN 0160-791X. https://doi.org/10.1016/j.techsoc.2020.101417.

Lob van Gennep. (2022). *Voorkeursbeslissing Lob van Gennep.* https://www.lobvangennep.nl/applications/lobvangennep/files/Downloads/voorkeursbeslissing/Voorkeursbeslissing-Lob-van-Gennep.pdf

Mackay, C. M., Schmitt, M. T., Lutz, A. E., & Mendel, J. (2021). Recent developments in the social identity approach to the psychology of climate change. *Current Opinion in Psychology, 42*, 95–101.

Marikyan, D., & Papagiannidis, S. (2023). Protection motivation theory: A review. In S. Papagiannidis (Ed.), *TheoryHub Book: This handbook is based on the online theory resource: TheoryHub* (pp. 78–93). https://open.ncl.ac.uk/theory-library/TheoryHubBook.pdf

Martinez, G., Costas, S., & Ferreira, Ó. (2020). The role of culture for coastal disaster risk reduction measures: Empirical evidence from northern and southern Europe. *Advances in Climate Change Research, 11*(4), 297–309.

Meyer, R., & Kunreuther, H. (2017). *The ostrich paradox: Why we underprepare for disasters.* University of Pennsylvania Press.

Morss, R. E., Cuite, C. L., & Demuth, J. L. (2024). What predicts hurricane evacuation decisions? The importance of efficacy beliefs, risk perceptions, and other factors. *NPJ Natural Hazards, 1*(1), 24. https://doi.org/10.1038/s44304-024-00025-8

Nasi, L., Jans, V., & Steg, L. (2024). Individual transilience in the face of the covid-19 pandemic. *Journal of Environmental Psychology, 93*, Article 102188.
Nationaal Deltaprogramma. (2024). *Advies Signaalgroep Deltaprogramma 2024*. https://www.deltaprogramma.nl/documenten/publicaties/2024/11/29/advies-signaalgroep-deltaprogramma-2024
Nationaal Deltaprogramma. (n.d.). *Deltabeslissing Waterveiligheid*. https://www.deltaprogramma.nl/themas/waterveiligheid/deltabeslissing
Nickerson, R. S. (1998). Confirmation Bias: A Ubiquitous Phenomenon in Many Guises. *Review of General Psychology, 2*(2), 175–220. https://doi.org/10.1037/1089-2680.2.2.175
NOS. (2024, November 3). *Spaanse koning bekogeld met modder, uitgejouwd bij bezoek aan rampgebied*. Nederlandse Omroep Stichting. https://nos.nl/artikel/2543066-spaanse-koning-bekogeld-met-modder-uitgejouwd-bij-bezoek-aan-rampgebied
Oakley, M., Mohun Himmelweit, S., Leinster, P., & Casado, M. R. (2020). Protection Motivation Theory: A Proposed Theoretical Extension and Moving beyond Rationality—The Case of Flooding. *Water, 12*(7), 1848. https://www.mdpi.com/2073-4441/12/7/1848
Palomo-Vélez, G., Perlaviciute, G., Contzen, N., & Steg, L. (2024). Trusting the minister or trusting the mayor? Perceived competence and integrity of central and local Dutch institutions governing energy matters. *Environmental Research Communications, 6*(4), Article 045009.
Parlementaire enquêtecommissie aardgaswinning Groningen. (2023). *Groningers boven gas*. https://www.tweedekamer.nl/Groningen/rapport
Committee of Inquiry on Childcare Benefits / Parlementaire ondervragingscommissie kinderopvangtoeslag. (2020). *Ongekend onrecht*. https://www.tweedekamer.nl/sites/default/files/atoms/files/20201217_eindverslag_parlementaire_ondervragings-commissie_kinderopvangtoeslag.pdf
Peters, L. E. (2021). Beyond disaster vulnerabilities: An empirical investigation of the causal pathways linking conflict to disaster risks. *International Journal of Disaster Risk Reduction, 55*, Article 102092.
Pidgeon, N., & Henwood, K. (2010). The Social Amplification of Risk Framework (SARF): Theory, Critiques, and Policy Implications. In P. Bennett, K. Calman, S. Curtis, & D. Fischbacher-Smith (Eds.), *Risk Communication and Public Health*. Oxford University Press. https://doi.org/10.1093/acprof:oso/9780199562848.003.04
Poortinga, W., Demski, C., & Steentjes, K. (2023). Generational differences in climate-related beliefs, risk perceptions and emotions in the UK. *Communications Earth & Environment, 4*(229), 1–8. https://doi.org/10.1038/s43247-023-00870-x
Poumadere, M., Mays, C., Le Mer, S., & Blong, R. (2005). The 2003 heat wave in France: Dangerous climate change here and now. *Risk Analysis: An International Journal, 25*(6), 1483–1494.
Renes, R. J. (2021). *De klimaatspagaat: over de psychologische uitdagingen van duurzaam gedrag*. Uitgeverij Creating Tomorrow.
Rijksoverheid. (n.d. a). *February 2020: Eerste coronabesmetting in Nederland*. https://www.rijksoverheid.nl/onderwerpen/coronavirus-tijdlijn/februari-2020-eerste-coronabesmetting-in-nederland
Rogers, R. W. (1975). A Protection Motivation Theory of Fear Appeals and Attitude Change1. *The Journal of Psychology, 91*(1), 93–114. https://doi.org/10.1080/00223980.1975.9915803
Rosenthal, U. (1988). Studies in Holland flood disaster 1953: An essay on the proto-sociology of disaster. *International Journal of Mass Emergencies & Disasters, 6*(3), 233–251.
RVO. (2025). *Zakendoen in Sint Maarten*. Rijksdienst voor Ondernemend Nederland. https://www.rvo.nl/onderwerpen/landen-en-gebieden/sint-maarten
Saul, J. (2022). *Collective trauma, collective healing: Promoting community resilience in the aftermath of disaster*. Routledge.
Schenk, G. J. (2015). 'Learning from History'?: Chances, problems and limits of learning from historical natural disasters. In *Cultures and Disasters* (pp. 72–87). Routledge.

Schwaller, N. L., Kelmenson, S., BenDor, T. K., & Spurlock, D. (2020). From abstract futures to concrete experiences: How does political ideology interact with threat perception to affect climate adaptation decisions? *Environmental Science & Policy, 112*, 440–452.

SCP. (2025). *Klimaat en Samenleving—Burgerperspectieven.* Sociaal en Cultureel Planbureau. https://www.scp.nl/publicaties/publicaties/2025/04/17/klimaat-en-samenleving

Slovic, P. (1992). *Perception of risk: Reflections on the psychometric paradigm.* Preager.

Snel, E., el Farisi, B., Engbersen, G., & Krouwel, A. (2022). Sociaaleconomische status en institutioneel vertrouwen in een tijd van corona. *Tijdschrift Sociologie, 3.*

Stark, A. (2018). *Public inquiries, policy learning, and the threat of future crises.* Oxford University Press.

Stuurgroep Lob van Gennep. (2020). *Notitie Reikwijdte en Detailniveau Lob van Gennep. Te onderzoeken alternatieven en effecten.* https://www.lobvangennep.nl/applications/lobvangennep/files/Downloads/nrd/NRD_LvG_20200129_definitief.pdf

Tenbrink, T., & Willcock, S. (2023). Place attachment and perception of climate change as a threat in rural and urban areas. *PLoS ONE, 18*(9), Article e0290354.

Terpstra, T. (2011). Emotions, Trust, and Perceived Risk: Affective and Cognitive Routes to Flood Preparedness Behavior. *Risk Analysis, 31*(10), 1658–1675. https://doi.org/10.1111/j.1539-6924.2011.01616.x

Tezar, T., & Setiadi, R. (2023). Risk Perception of Small Islands Community on Climate Change: Evidence From Mepar and Baran Islands Indonesia. *Island Studies Journal, 19*(1), 1–21.

Thaler, R. H., & Shefrin, H. M. (1981). An Economic Theory of Self-Control. *Journal of Political Economy, 89*(2), 392–406. http://www.jstor.org/stable/1833317

Thomalla, F., Smith, R., & Schipper, E. L. F. (2015). Cultural Aspects of Risk to Environmental Changes and Hazards. In *Disaster's impact on livelihood and cultural survival: Losses, opportunities, and mitigation*, Routledge.

Thomas, W. I. (1928). The behavior pattern and the situation. *Publications of the American Sociological Society, 22*, 1–14.

UNICEF. (2024). *Beat the heat: Child health amid heatwaves in Europe and Central Asia.* https://www.unicef.org/eca/reports/beat-heat-2024

Universitat Autònoma de Barcelona. (n.d.). *Beschrijving case study Itteren en Borharen in de gemeente Maastricht.* https://webs.uab.cat/capflo/case-studies/the-netherlands-dutch/

Van Buuren, A. (2019). *The Dutch delta approach: The successful reinvention of a policy success.* Oxford University Press. https://doi.org/10.1093/oso/9780198843719.003.0011

Van den Brink, M. (2021). Rijkswaterstaat: Guardian of the Dutch delta. In A. Boin, L. A. Fahy, & P. 't Hart (Eds.), *Guardians of Public Value: How Public Organisations Become and Remain Institutions* (pp. 237–261).

Van der Laan, L. N., van der Waal, N. E., & De Wit, J. (2022). *Eindrapportage Corona Melder Evaluatie-Survey LISS panel wave 6.* Tilburg University.

Van der Linden, S. (2017). *Determinants and measurement of climate change risk perception, worry, and concern.* Oxford University Press.

Van der Sanden, E. (2019). Anderhalf jaar in Nederland na orkaan Irma: een leven met financiële problemen, vooroordelen en trauma's. *Caribisch Netwerk.* https://land-na-orkaan-irma-een-leven-met-financiele-problemen-vooroordelen-en-traumas/

Van Valkengoed, A. M., & Steg, L. (2019a). Meta-analyses of factors motivating climate change adaptation behaviour. *Nature Climate Change, 9*(2), 158–163.

Van Valkengoed, A. M., & Steg, L. (2019b). *The psychology of climate change adaptation.* Cambridge University Press.

Van Valkengoed, A. M., Perlaviciute, G., & Steg, L. (2024). From believing in climate change to adapting to climate change: The role of risk perception and efficacy beliefs. *Risk Analysis, 44*(3), 553–565. https://doi.org/10.1111/risa.14193

Werkgroep Dorpsraden. (2024). *Kunnen de gevolgen van een overstroming in de Lob van Gennep worden beperkt?* https://maasburen.nl/files/maasburen.nl/uploads/files/740811/Nota%20Overstromingsbestendig%20maken%20Lob%20v%20G%20jan2024.pdf

Wijkhuis et al. (2018). *Orkaan Irma treft Sint Maarten en Caribisch Nederland. Een evaluatie van het door hetministerie van IenW geleverde crisismanagement*. IFV Instituut Fysieke Veiligheid.
Witte, K. (1992). Putting the fear back into fear appeals: The extended parallel process model. *Communication Monographs, 59*(4), 329–349. https://doi.org/10.1080/03637759209376276
WRR. (2014). *Met kennis van gedrag beleid maken* (rapport nr.92). Wetenschappelijke Raad voor het Regeringsbeleid. https://www.wrr.nl/publicaties/rapporten/2014/09/10/met-kennis-van-gedrag-beleid-maken
WRR. (2022). *Coronascenario's doordacht: Handreiking voor noodzakelijke keuzes*. Wetenschappelijke Raad voor het Regeringsbeleid. https://www.wrr.nl/publicaties/publicaties/2022/09/05/coronascenarios-doordacht-handreiking-voor-noodzakelijke-keuzes
WRR. (2023). *Grip: Het maatschappelijk belang van persoonlijke controle* (rapport nr.108). Wetenschappelijke Raad voor het Regeringsbeleid. https://www.wrr.nl/publicaties/rapporten/2023/11/30/grip

Chapter 4
Social Cohesion and Climate Events

Contents

Key Points

- Social cohesion refers to the degree of connectedness within a society. When a society exhibits a high level of social cohesion, citizens are able to live and work together harmoniously.
- Social cohesion influences how people act before, during and after a climate event. A strong social safety net can make the difference between life and death during a climate event. Recovery is often more successful within close-knit communities.
- A climate event can have repercussions for social cohesion. Adversity often leads to temporary solidarity, but the long-term impact is not always positive. The way in which the authorities respond to a climate event is crucial in this regard.

4.1 Offering a Helping Hand is not Enough

> Immediately after the disaster, many parties came forward to help and repair the damage. The Netherlands is good at crisis management: helping people, addressing the initial emergency, but then it stalls. As a municipality, we moved on to the next phase – the aftermath – quite quickly, but now, nearly a year later, that's still where we are. (…) This morning, a member

R. El-Dardiry et al., *People and Climate*, Research for Policy,
https://doi.org/10.1007/978-3-032-25831-1_4

> of our field staff members came into my office. Right after the flood, he left the mess at home to help other affected residents. Now he's struggling and deeply worried about his finances, and only now does he dare to ask for help. Eleven months have passed. For that reason alone, I keep insisting that something must change in this country.[1]
>
> Daan Prevoo, Mayor of Valkenburg (province of Limburg)

On 14 and 15 July 2021, the river Geul overflowed its banks in the far south of the Netherlands. This flood caused damage worth hundreds of millions of euros in the municipality of Valkenburg. People were forced to leave their homes and business owners had to shut down their operations. In the quote above, the mayor of Valkenburg notes that help was offered from all sides in the immediate aftermath of the disaster. In the previous chapter, we already saw that a disaster can lead to greater solidarity among people.[2] The mayor explains that one member of his field staff found himself in a precarious situation nearly a year after the disaster. This man gave a helping hand to the community, but did not dare to ask for help himself. A social safety net is therefore far from self-evident, even for those who contribute to the collective.

The flooding of the Geul was the result of excessive rainfall in the summer—one of the weather phenomena that, according to the Royal Netherlands Meteorological Institute, will occur more frequently due to climate change, along with more heat, drought and rainier winters. These climate events have an impact on social cohesion. Equally importantly, social cohesion is one of the factors that determines whether a society can successfully cope with changes in its weather.

In this chapter, we examine the interaction between social cohesion and the consequences of climate change. We begin with a brief explanation of what we mean by social cohesion (Sect. 4.2). We then discuss how social cohesion can help to address the consequences of climate change (Sect. 4.3) and how climate change itself affects social cohesion (Sect. 4.4). Illustration 4.1 shows how social cohesion and the impact of a climate event, such as a flooding caused by rainfall, are interrelated.

4.2 Social Cohesion: the Whole is Greater than the Sum of its Parts

What holds a society together? Many scholars have struggled with this question. And it is not only academics who are interested in it: a society falling apart is a frightening prospect for many people. Concerns about social divides and increasing segregation are voiced frequently.[3]

By 'social cohesion', we refer to the connectedness within society: the social fabric that links individuals and enables citizens to live and work together.[4] The concept has various definitions, partly because social cohesion is studied across multiple

---

[1] Dutch Association of Insurers (2022).

[2] Cf. Meijer (2022).

[3] Tiemeijer (2017), SCP (2024b).

[4] Glas and Mulder (2018). In this context, Van Houwelingen (2016) also refers to an 'invisible glue'.

Illustration 4.1 Social cohesion

academic disciplines. More important than a precise definition are the themes that emerge when we reflect on the connectedness within society.[5]

Interactions between people play a central role: with whom do people have contact, and what is the nature of these contacts? We can view society as a web of networks that connect people (Fig. 4.1). At the most basic level, we find networks of people who have much in common (*bonding*). The mutual trust among members of such a network is high, and cooperation is almost taken as self-evident. Examples include close-knit families, neighbourhood communities or student societies. In addition, people seek contact with others who are less like themselves but with whom they associate in other ways. This gives rise to connections between the more cohesive networks (*bridging*). Many of these connections arise from shared activities. Sports clubs, political parties and workplaces are typical examples where such contacts are established. A third type of connection arises when bridging occurs between people with very different positions of social power (*linking*). This can happen, for example, when someone meets a local councillor through the football club.[6]

These three types of connections are generally referred to in the literature as different forms of social capital.[7] For social cohesion in society as a whole, the interplay between these connections is important. For example, social problems can arise when people from different ethnic or socio-economic backgrounds rarely encounter each other in a positive way. In the worst case, there is a risk that members of different groups will approach each other with hostility. Although people in such a situation may feel strongly connected to a small group, cohesion at the societal level remains low.

Political scientist Robert Putnam established a link between social capital and trust in the authorities and politics.[8] By joining clubs and engaging in volunteer work, people are exposed to different perspectives and experience first-hand how fruitful cooperation can be achieved. In this way, citizens effectively train their capacity to listen and develop an understanding of difficult trade-offs. As a result, citizens may also develop greater trust in politics. After all, political decision-making is often a slow process in which understanding other viewpoints is essential to reaching compromises.[9] In the Netherlands, the magnitude of this effect appears to be relatively limited.[10]

---

[5] For an overview of various scientific definitions of social cohesion, see Appendix B to SCP (2023); see also Brock et al. (2019).

[6] Szreter (2002); Putnam (2000).

[7] Researchers at the SCP define social capital as 'the total resources available to a person or group through their social network, which can help them achieve goals that would otherwise not be possible or only at (much) higher cost'. Regarding the difference between social cohesion and social capital, they note: 'Where social cohesion concerns a "state of affairs" regarding the quality of social relationships and the coherence between various social groups, social capital is (...) to be viewed as a resource from which individuals or social groups draw to enhance their agency.' (SCP 2023: 31; see also Dijkstra 2012: 14). Translated by the WRR.

[8] Putnam (1993).

[9] Korsten and De Goede (2006).

[10] Dekker and Van der Meer (2011).

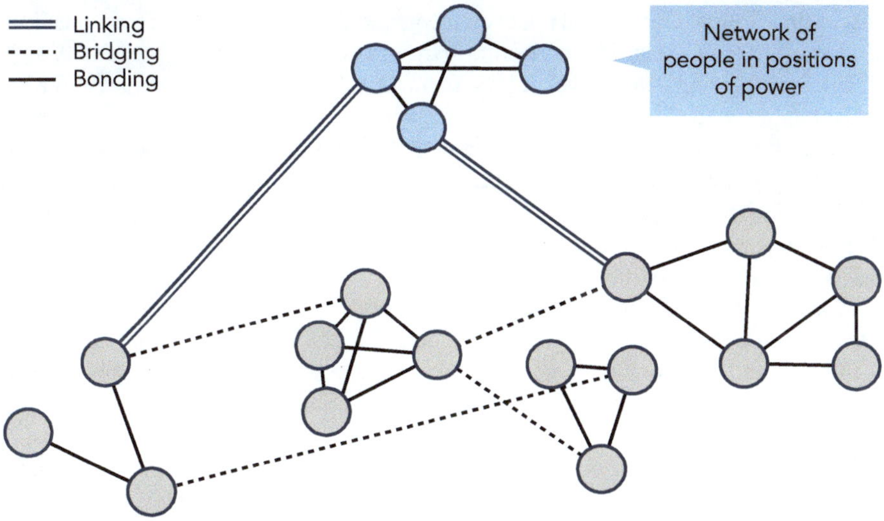

Fig. 4.1 Social connections in society: bonding, bridging and linking[11]

Chan et al. distinguished three criteria that a society with a high degree of social cohesion must meet. First, people in such a society are able to trust, help and cooperate with each other. Second, they share a common identity and feel connected to one another. Finally, these feelings lead to concrete actions. In other words, social cohesion is also expressed in behaviour.[12]

Much scientific research seeks to map out social cohesion in societies. The three criteria mentioned above provide useful guidelines for this. Many researchers focus on trust and participation.[13] Among other things, they do so by conducting surveys in which they ask people about their trust in others, their willingness to help others or their membership of civil-society organisations.[14] Trust in public figures and institutions is also often included.[15] At best, studies like these provide an indication of the level of social cohesion because of the risk that respondents give socially desirable answers or that connectedness manifests itself in other ways.[16]

If we look at such indicators of social cohesion in the Netherlands, we see a mixed picture. Within Europe, the Netherlands is one of the countries with the highest percentage of volunteers. Nearly half of the Dutch population engages in volunteer work each year.[17] Compared to other countries, the Dutch have a high level of

[11] Cf. Aldrich (2012a).

[12] Chan et al. (2006).

[13] Schmeets and Exel (2022).

[14] CBS (2018).

[15] SCP (n.d.)

[16] Van Houwelingen (2016).

[17] Schmeets et al. (2016); CBS (2024a).

trust in their fellow citizens.[18] However, the government response to the coronavirus pandemic has led to a significant decline in trust in the authorities: in April 2020, 69% of Dutch people had a high level of trust in the Dutch government, compared to 22% in September 2023.[19] There have been further developments indicating declining cohesion. For example, contact between neighbours is decreasing. Whereas 67% of Dutch people aged 45 to 55 reported having daily or weekly contact with neighbours in 2012, this had dropped to 58% by 2023. Other age groups show a similar trend.[20] More young people aged 15 to 25 reported feeling socially lonely in 2023 compared to 2019 (11.9% versus 10.1%).[21]

Moreover, there are groups who feel excluded from the rest of society.[22] Recent research by the Netherlands Institute for Social Research has shown that the worlds of the rich and the poor have become increasingly separated.[23] In highly diverse neighbourhoods, social interaction is more complex. This is sometimes associated with reduced cohesion.[24]

Urbanisation also plays a role in the level of social cohesion people experience. According to the 2023 Safety Monitor compiled by Statistics Netherlands, people living in rural areas experience more social cohesion than residents of large cities. They feel more connected to their neighbours, know each other better and are more willing to help one another.[25]

4.3 The Role of Social Cohesion in a Changing Climate

A recent report on the evacuation behaviour of victims during the flooding in Valkenburg repeatedly emphasised the importance of a strong social network. A circle of friends, family and acquaintances increases the likelihood that someone will be informed in time about a disaster. Moreover, members of the social network often lend a helping hand. In the absence of private transport, a lift from an acquaintance can make all the difference.[26] Multiple empirical studies have shown that communities with relatively high levels of cohesion recover more quickly after a disaster.[27] This was also the case in Valkenburg:

---

[18] CBS (2022, 2024e).

[19] Engbersen (2022); Snel et al. (2024).

[20] CBS (2023, 2024b).

[21] CBS (2024c).

[22] CBS (2020).

[23] SCP (2024b).

[24] WRR (2018).

[25] CBS (2024d).

[26] Van den Berg et al. (2022); Putnam (2000).

[27] Townshend et al. (2014); Ntontis et al. (2020); Reckin and Weil (2015).

> Especially when it came to cleaning up and subsequent renovations, the social network played a major role for many people. One victim who had received a lot of help from friends and family said: "If we hadn't had that (...), I don't know how we would have managed. We could never have done it alone. (...) If you don't have a social network because of who you are, that's really sad in those circumstances." A weaker network can therefore mean that the consequences of the flooding are more severe and last longer: "There are still elderly people living alone who had to organise everything themselves, whose homes have not yet been repaired and who are still living upstairs."[28]

The example of the municipality of Valkenburg shows that social cohesion is important both during and after a climate event. Interactions between people also play a key role in taking protective climate adaptation measures—that is, *even before* climate threats have led to problems. In any case, while some climate events may not result in major catastrophes as in Valkenburg, they can still cause significant disruption to social life. In such cases, adaptation can help to reduce social damage. For all these stages—before, during and after—we discuss how social cohesion plays a role.

4.3.1 Social Cohesion and Adaptive Action

In 2018 and 2019, an international team of scientists, including researchers from Wageningen University & Research, conducted studies in the province of Groningen, in the north-east of the Netherlands.[29] They wanted to determine to what extent different types of connections—bonding, bridging and linking—are associated with greater resilience among farmers. To investigate this, they conducted interviews with around a dozen farmers and analysed survey results from more than 70 farmers. They also organised a panel discussion and spoke with agricultural experts. They discovered that having a combination of different forms of social capital was linked to greater resilience. Farmers whose social networks were largely limited to similar people (bonding) mainly focused on strengthening existing production processes. Farmers with broader social networks—in other words, networks characterised by bridging and linking—were more likely to enhance their resilience by exploring alternative business models. Through contacts outside their own circle, they learned more about cultivating different crops or how a bed and breakfast could provide additional income.

The above study shows that social capital is important when implementing changes or taking precautionary measures. This is equally true in the context of climate adaptation. Close-knit communities with relatively little input from outside may struggle to adapt to a changing climate. Broadly speaking, there are two causes for this. Firstly, a close-knit community characterised by strong bonding may have a knowledge gap. Due to a lack of contacts outside their own network, community members can find it difficult to overcome any information deficit. In this sense, a lack of curiosity can hinder such a community. Secondly, the social norms within a group

[28] Van den Berg et al., (2022: 33).

[29] Slijper et al. (2022).

can resist change. For example, this may be the case if members of a community derive their identity from a particular way of life or a very specific place (see also Chapter 3).[30]

At the same time, close-knit communities offer opportunities. The high level of trust among members of such a community means they are sometimes willing to set aside their own interests for the greater good.[31] For example, American research has shown that the social capital present in such a community encourages its members to make greater efforts to reduce the risks of wildfires.[32] For many activities aimed at increasing climate resilience, however, all forms of social connections are important. For instance, successful citizen initiatives (see Box 4.1) require careful coordination with the municipality and benefit from knowledge exchange with initiators in other locations.[33]

Box 4.1 Social cohesion and citizen initiatives

Across the Netherlands, people are launching 'citizen initiatives' with the aim of making their street or neighbourhood more climate-resilient. These initiatives range from installing green roofs and selling rainwater tanks to improving home insulation. For example, residents of the Kattenbosch neighbourhood in Rosmalen, in the south of the Netherlands, have ensured that dozens of roofs in their area are now covered with stonecrops. As a result, the homes are better protected against heat and extreme rainfall.[34] In the city of Tilburg, a citizen initiative aims to create 1000 so-called water-friendly gardens within three years. This will reduce the risk of flooding.[35] In the city of Zwolle, there are several citizen initiatives focused on making homes more sustainable. People jointly purchase insulation materials or ensure that neighbours have easy access to expertise.[36] As these examples show, citizen collectives make an important contribution to climate adaptation in numerous ways. However, this does not mean that setting up a citizen initiative is easy. Funding and objections from the authorities can sometimes be significant obstacles.[37] Successful citizen initiatives often bring together different types of social connections: bonding is important at the neighbourhood level, while bridging and linking play a role in contact with officials or initiators elsewhere.

[30] Smith et al. (2012).

[31] Ostrom and Ahn (2009); Ireland and Thomalla (2011).

[32] Bihari and Ryan (2012); Agrawal and Monroe (2006).

[33] Kleinhans (2017).

[34] Provincie Noord-Brabant. (2018).

[35] Huisje Boompje Beter (n.d.).

[36] Platform duurzaam Zwolle (n.d.).

[37] Oostveen (2023).

4.3.2 The Importance of Social Cohesion During Climate Events

As discussed in Chapter 1, the consequences of climate change, such as prolonged heat and extreme rainfall, can be socially disruptive. The previously mentioned flooding of the river Geul is an example of this. At the time, the Dutch government classified this flood as a disaster.[38] During a crisis, social networks are often crucial in preventing or mitigating suffering.

The role of social cohesion is extensively described in disaster literature. Disasters often lead to altruism: in a crisis situation, people are more willing to help each other. It is perhaps not surprising that people draw on their social capital when other forms of capital are under severe pressure. The late sociologist Russell Dynes commented on this:

> 'Most importantly, of all the forms of capital, [social capital] is less damaged and less affected. Consequently, during the emergency period, it is the form of capital that serves as the primary base for a community response. In addition, social capital is the only form of capital which is renewed and enhanced during the emergency period.'[39]

In an emergency, a strong social network can quite literally mean the difference between life and death. Conversely, people who are socially isolated are less likely to survive a catastrophe. They lack the contacts who look out for them, provide information or help with evacuation if necessary.[40]

In his book *Heat wave*, the sociologist Eric Klinenberg demonstrates how important social cohesion is in coping with heat in an urban environment. He reached this conclusion after conducting extensive research on a heatwave that struck Chicago in 1995 and caused hundreds of deaths within a short period of time. The quality of what he would later call 'social infrastructure'—'the physical places and organizations that shape the way people interact'[41]—proved to be one of the main explanatory factors for excess mortality in a given neighbourhood.[42] Older people who chatted regularly with neighbours and were therefore less likely to become isolated had a greater chance of surviving the heatwave. The design of the spatial environment can encourage such social interactions. Consider, for example, public spaces such as libraries or parks. In Box 4.2, we discuss Klinenberg's study in more detail.

---

[38] NOS (2021).

[39] Dynes (2003: 9).

[40] Aldrich and Meyer (2015).

[41] Klinenberg (2018: 5); see also Kolner and Engbersen (2024).

[42] Klinenberg (2015).

Box 4.2 The importance of social infrastructure during a heatwave

In July 1995, the city of Chicago was struck by a heatwave. For several consecutive days, temperatures exceeded 40 °C, accompanied by high humidity. The hot air barely moved and smog descended down to street level. Particularly in the central parts of the city, an urban heat island effect developed: the many buildings and roads trapped the heat and reflected it, limiting night-time cooling. The excess mortality recorded during the four most critical days amounted to 739 people.

The municipality was forced to deploy seven refrigerated trucks to handle the enormous influx of bodies. Images of this, broadcasted from media helicopters, left a deep impression on the city's collective psyche. Forty-one deceased individuals ended up in a mass grave because their bodies remained unclaimed. Most victims were elderly people living in poverty. Many had no air conditioning or could not afford to turn it on. There were also power outages that made cooling systems unusable. A disproportionately high number of victims lived in neighbourhoods plagued by high crime rates and from which government agencies and aid organisations had effectively withdrawn for much of the daytime. As a result, residents did not dare to sleep with open windows or outdoors.

A 'social autopsy' of the disaster revealed that this pattern of victimhood was the result of a combination of age, socio-economic and social cohesion factors.[43] Virtually all victims were elderly, lived in poverty and lived alone. More men than women died. On average, women are more socially connected to family members and people in their community than men. The vast majority of victims lived in African American neighbourhoods; in Latino and white neighbourhoods with similar socio-economic conditions, excess mortality was significantly lower. The underlying reason for this was that family and community structures were often much more disintegrated in the African American neighbourhoods, crime rates were much higher and institutions were less active. As a result, vulnerable individuals were more often overlooked, let alone helped.

Evacuations are particularly stressful for people and can lead to mental health problems. It is precisely at such times that a social network can ensure that victims do not feel abandoned. Social cohesion increases the likelihood that victims have sufficient access to resources and other facilities.[44] Social connections provide people with an additional layer of protection. For example, Dutch people with high social capital experienced less fear and stress during the coronavirus pandemic.[45]

---

43 Klinenberg (2002).

44 Greene et al. (2015).

45 Snel et al. (2022).

However, this does not mean that greater social cohesion during a crisis *always* leads to better outcomes. Although strong social ties can ensure that people are better prepared and evacuate more quickly, cohesion can also lead to risky behaviour.[46] For instance, individuals who feel a strong connection to a particular group may be inclined to try to rescue others, thereby putting themselves in danger.

4.3.3 Social Cohesion and Recovery

Two months after the North Sea flood of 1953, female volunteers from all over the Netherlands travelled to the affected areas in the provinces of Zeeland, South Holland and North Brabant.[47] Upon arrival, they received a headscarf and overalls. Armed with brooms and buckets, they set to work cleaning the thick layer of mud from floors and furniture in homes. Scrubbing, scouring, mopping: in two days, these 'mud girls' made a house at least somewhat habitable again (Fig. 4.2). The gratitude of the local population was immense, as is evident from a fragment of a song of praise sung in the village of Oude-Tonge for several volunteers from the northern city of Haarlem:

> We give you thanks, dear Haarlem maids,
> for coming to help us cope.
> Your achievements need no further praise –
> they restored our spirits and hope.[48]

After a disaster, a period usually follows in which attempts are made to return to the previous situation. Depending on the scale and nature of the event, this recovery can require considerable time, space and energy. Moreover, mud girls do not always appear. When adrenaline levels drop and media attention wanes, victims may feel as though no one is looking out for them. Under such circumstances, feelings of resentment can arise easily—as described in the previous chapter.

Political scientist Daniel Aldrich identifies three reasons why social capital, even more than economic and human capital or the extent of the damage, is the most important predictor of successful recovery.[49] First, people can use a strong social network to exchange information and provide each other with financial and administrative support. Second, well-connected communities have greater organisational capacity. Collective action makes it easier to articulate shared needs and secure resources from

[46] Meyer (2018).

[47] The North Sea flood of 1953 was a catastrophic flood that occurred during the night of 31 January to 1 February 1953, caused by a severe north-westerly storm combined with a spring tide. This combination resulted in extremely high water levels, causing dykes to break in the provinces of Zeeland and North Brabant and on the South Holland islands. Large areas were inundated and 1836 people lost their lives. This disaster prompted the construction of the Delta Works, a series of dams and storm surge barriers to protect the Netherlands from flooding. The North Sea flood of 1953 remains deeply etched in the memory of many Dutch people.

[48] See, among others, Both (n.d.); Terhoeve (2023). Translated by the WRR.

[49] Aldrich (2012b).

Fig. 4.2 Mud girls. *Source* National Archives

the authorities. Neighbours who keep an eye out also help reduce the likelihood of looting. Third, social capital creates a barrier to leaving a devastated neighbourhood. People are less likely to turn their backs on a neighbourhood where they have an extensive social network; that capital is valuable. Community spirit thus provides an incentive for reconstruction.

4.4 Consequences of Climate Change for Social Cohesion

As in other languages, temperature is often associated in Dutch with emotions and behaviour. Someone who keeps a cool head remains calm; with a hot-tempered person, on the other hand, it is best to keep your distance until they have cooled down. These associations are not entirely unfounded. A Dutch group of scientists placed people in a warm bath and then examined how the higher temperature influenced their decision-making. Their results show that people with a higher body temperature spend less time making choices. Tempers literally run shorter at higher temperatures, although there is still much scientific debate about the precise impact of heat on cognitive ability.[50] Other studies show that increasing heat is associated

[50] For a brief video presentation on why people tend to be more aggressive in summer, see University of the Netherlands (2022). For scientific publications on heat and behaviour, see Van Maanen et al. (2019); Zhang et al. (2019).

with more aggression in society.[51] For example, domestic violence increases during heatwaves.[52] Now that the likelihood of heatwaves in our country is increasing, we must therefore take into account a rise in violent incidents.[53]

The risk that the consequences of climate change will worsen interactions between people is not limited to heat. Reduced cohesion can also arise when residents of an affected neighbourhood decide not to return after an evacuation. Houses may then remain vacant or be occupied by many new residents. The original inhabitants who do not leave may consequently feel out of place.[54] And when recovery efforts after a climate event are insufficient, there is a threat of deterioration, which in turn reduces social cohesion over time.

At the same time, as we saw in the introduction to this chapter and in the previous chapter, disasters can also lead to greater solidarity and a strengthening of identity. Especially during the recovery phase, new connections can emerge.[55] In the Roombeek district of Enschede, the municipality deliberately promoted broad resident participation during the reconstruction phase after the fireworks disaster.[56] Many victims of the fireworks disaster therefore felt taken seriously and proud that they were able to contribute to the recovery effort.[57]

Finally, it should be noted that the way in which the authorities design climate adaptation measures and respond to crisis situations can have a direct impact on social cohesion within a society.[58] Across the country, squares and streets are being greened. More greenery in the neighbourhood not only helps to cope with heat and excessive rainfall, but can also strengthen the local social infrastructure, because residents are more likely to meet outdoors when the environment is inviting. In addition, cohesion can be reinforced by a strong national identity. For example, the Dutch are generally proud of the Delta Works.[59] They have become a symbol of our struggle against water.

The societal consequences of the coronavirus measures are quite different. Trust in the Dutch government has declined sharply. And although neighbourhood relations and mutual trust did not change significantly during and after the pandemic, citizens are concerned about increasing polarisation. Supporters and opponents of the government's measures find it difficult to engage in dialogue.[60] It is certainly not inconceivable that climate adaptation measures or the official response to climate disasters could lead to similar polarisation in the future. A disaster can drive people

[51] Anderson (2001); Anderson et al. (1997); Williams et al. (2015); Miles-Novelo and Anderson (2019).

[52] Sanz-Barbero et al. (2018).

[53] National Institute for Public Health and the Environment (2024); Park (2024).

[54] Van Holm et al. (2018).

[55] Meyer (2018).

[56] Denters and Klok (2010).

[57] Bazen (2014: 172).

[58] Carlin et al. (2014).

[59] Jensen (2020).

[60] Snel et al. (2023).

apart if the eventual response proves unsatisfactory, if authorities fail to provide sufficient accountability afterwards or if the suffering of those affected is not adequately acknowledged and remembered.[61]

4.5 Conclusion

The saying 'together we are strong' also applies when considering the societal impacts of climate change. People who exchange information learn more quickly which measures can be effective in dealing with extreme rainfall, drought and heat. In times of great stress, such as during a flood, a social network serves as a safety net. People who know and trust each other are more likely to offer help. Finally, social relationships are important for recovery after a disaster.

Nevertheless, it would be too simplistic to view social cohesion solely as a positive force. In this chapter, we have also encountered examples showing how close-knit communities can hinder the development of a climate-resilient society as a whole.

The relationship between climate change and social cohesion also goes both ways. Changing weather patterns and climate-related events influence social cohesion within society.

In the previous chapter, we saw that the way the authorities deal with climate events affects the mentality of individuals and communities. The same applies to social cohesion. Trust in institutions is one of the factors that plays a role here. Government policy can facilitate the formation of connections between people. High-quality public spaces and amenities such as libraries enable people to meet one another.

A policy that broadly strengthens social cohesion can reduce climate vulnerability. This potential of social cohesion is particularly important in climate-vulnerable parts of society. Social cohesion forms a buffer that can help prevent a weak socio-economic position from deteriorating further in times of crisis.

References

Agrawal, S., & Monroe, M. C. (2006). Using and improving social capital to increase community preparedness for wildfire. In USDA, *The Public and Wildland Fire Management: Social Science Findings for Managers* (pp. 163–167). US Department of Agriculture, Forest Service.

Aldrich, D. P. (2012a). *Building resilience: Social capital in post-disaster recovery*. University of Chicago Press.

Aldrich, D. P. (2012b). Social, not physical, infrastructure: The critical role of civil society after the 1923 Tokyo earthquake. *Disasters*, *36*(3), 398–419. https://doi.org/10.1111/j.1467-7717.2011.01263.x

[61] Bos et al. (2005).

Aldrich, D. P., & Meyer, M. A. (2015). Social capital and community resilience. *American Behavioral Scientist, 59*(2), 254–269.

Anderson, C. A. (2001). Heat and Violence. *Current Directions in Psychological Science, 10*(1), 33–38. https://doi.org/10.1111/1467-8721.00109

Anderson, C. A., Bushman, B. J., & Groom, R. W. (1997). Hot years and serious and deadly assault: Empirical tests of the heat hypothesis. *Journal of Personality and Social Psychology, 73*(6), 1213–1223. https://doi.org/10.1037//0022-3514.73.6.1213

Bazen, J. C. (2014). *We designed this neighborhood! Neighborhood satisfaction of local non-gentrifying residents with the reconstructed Roombeek neighborhood in Enschede, The Netherlands.* Utrecht University. https://studenttheses.uu.nl/handle/20.500.12932/17837

Bihari, M., & Ryan, R. (2012). Influence of social capital on community preparedness for wildfires. *Landscape and Urban Planning, 106*(3), 253–261. https://doi.org/10.1016/j.landurbplan.2012.03.011

Bos, C. K., Ullberg, S., & 't Hart, P. (2005). The Long Shadow of Disaster: Memory and Politics in Holland and Sweden. *International Journal of Mass Emergencies & Disasters, 23*(1), 5–26. https://doi.org/10.1177/028072700502300101

Both, J. (n.d.). *De Wederopbouw op Goeree-Overflakkee.* https://geschiedenisvanzuidholland.nl/verhalen/verhalen/de-wederopbouw-op-goeree-overflakkee/

Brock, A., Kwakernaak, M., de Meere, F., & Boutellier, H. (2019). *Literatuurstudie Sociale Cohesie. Ten behoeve van de ontwikkeling van de Maatschappelijke Impact Analyseren (MIA).* Verwey Jonkers Instituut.

Carlin, R. E., Love, G. J., & Zechmeister, E. J. (2014). Trust Shaken: Earthquake Damage, State Capacity, and Interpersonal Trust in Comparative Perspective. *Comparative Politics, 46*(4), 419–437. http://www.jstor.org/stable/43664117

CBS. (2018). *Sociale cohesie (binding met de buurt).* Centraal Bureau voor de Statistiek. https://www.cbs.nl/nl-nl/nieuws/2018/01/tevredener-met-het-leven-bij-sterkere-binding-met-buurt/sociale-cohesie—binding-met-de-buurt

CBS. (2020). *Sociale uitsluiting in Nederland: wie staat aan de kant*? Centraal Bureau voor de Statistiek. https://www.cbs.nl/nl-nl/longread/statistische-trends/2020/sociale-uitsluiting-in-nederland-wie-staat-aan-de-kant-/1-inleiding

CBS. (2022). *Vertrouwen in Tweede Kamer steeg in 2020 en daalde in 2021.* Centraal Bureau voor de Statistiek. https://www.cbs.nl/nl-nl/nieuws/2022/13/vertrouwen-in-tweede-kamer-steeg-in-2020-en-daalde-in-2021

CBS. (2023). *Bijna 70 procent van ouderen heeft minstens elke week contact met de buren.* Centraal Bureau voor de Statistiek. https://www.cbs.nl/nl-nl/nieuws/2023/38/bijna-70-procent-van-ouderen-heeft-minstens-elke-week-contact-met-de-buren

CBS. (2024a). *Vrijwilligerswerk 2023.* Centraal Bureau voor de Statistiek. https://www.cbs.nl/nl-nl/longread/rapportages/2024/vrijwilligerswerk-2023?onepage=true#c-1—Inleiding

CBS. (2024b). *Sociale contacten en maatschappelijke participatie.* Centraal Bureau voor de Statistiek. https://opendata.cbs.nl/#/CBS/nl/dataset/85541NED/table?dl=922F0

CBS. (2024c). *Eenzaamheid; persoonskenmerken.* Centraal Bureau voor de Statistiek. https://opendata.cbs.nl/statline/#/CBS/nl/dataset/85766NED/table?ts=1731315091334

CBS. (2024d). *Veiligheidsmonitor 2023.* Centraal Bureau voor de Statistiek. https://www.cbs.nl/nl-nl/longread/rapportages/2024/veiligheidsmonitor-2023/1-inleiding-en-samenvatting

CBS. (2024e). *Vertrouwen in mensen, recht en politiek; Europese vergelijking.* Centraal Bureau voor de Statistiek. https://www.cbs.nl/nl-nl/cijfers/detail/80518ned?dl=1EB57

Chan, J., To, H.-P., & Chan, E. (2006). Reconsidering social cohesion: developing a definition and analytical framework for empirical research. *Social Indicators Research, 75*(2), 273–302. https://doi.org/10.1007/s11205-005-2118-1

Dekker, P., & Van der Meer, T. (2011). Geen vertrouwen meer? In L. C. J. M. Halman & I. J. P. Sieben (Eds.), *Respect man! Tolerantie, solidariteit en andere moderne waarden* (pp. 197–210). Celsus.

Denters, B., & Klok, P.-J. (2010). Rebuilding roombeek: Patterns of citizen participation in urban governance. *Urban Affairs Review, 45*(5), 583–607. https://doi.org/10.1177/1078087409356756

Dijkstra, A. (2012). *Sociale opbrengsten van onderwijs*. Vossiuspers UvA. https://pure.uva.nl/ws/files/1615318/122688_PDF_7303Dijkstra_webversie.pdf

Dynes, R. (2003). Finding order in disorder: Continuities in the 9–11 response. *International Journal of Mass Emergencies & Disasters, 21*(3), 9–23.

Engbersen, G. (2022). De maatschappelijke impact van covid-19—Sociologie van een pandemie. *De Sociologie en de Pandemie: Inzichten en Vooruitblik na Twee Jaar Coronacrisis*. https://pure.uvt.nl/ws/files/65910933/Sociologie_en_de_pandemie_compleet_2022_pdf

Glas, I., & Mulder, L. (2018). Diversiteit naar herkomst en sociale cohesie. *De nieuwe verscheidenheid, toenemende diversiteit naar herkomst in Nederland*. In R. Jennissen, G. Engbersen, M. Bokhorst & M. Bovens. WRR Verkenning nr 38.

Greene, G., Paranjothy, S., & Palmer, S. R. (2015). Resilience and vulnerability to the psychological harm from flooding: The role of social cohesion. *American Journal of Public Health, 105*(9), 1792–1795. https://doi.org/10.2105/ajph.2015.302709

Huisje Boompje Beter. (n.d.). *Blauwe tuinen zijn 'toffe tuinen'*. https://www.huisjeboompjebeter.nl/tilburg/blauwe-tuinen/

Ireland, P., & Thomalla, F. (2011). The role of collective action in enhancing communities' adaptive capacity to environmental risk: an exploration of two case studies from Asia. *PLoS Curr, 3*, Rrn1279. https://pubmed.ncbi.nlm.nih.gov/22045442/

Jensen, L. (2020). De strijd tegen het water in woord en beeld. Verbindingen tussen heden en verleden. In L. Jensen, & A. Duiveman (Eds.), *Welke verhalen vertellen we? Narratieve strategieën rondom water- beheer en zeespiegelstijging* (pp. 6–13). Radboud Universiteit. https://repository.ubn.ru.nl/handle/2066/227293

Jensen, L. (2024). *Rampen: Een nieuwe geschiedenis van Nederland*. Uitgeverij Prometheus.

Kleinhans, R. (2017). False promises of co-production in neighbourhood regeneration: The case of Dutch community enterprises. *Public Management Review, 19*(10), 1500–1518. https://doi.org/10.1080/14719037.2017.1287941

Klinenberg, E. (2002). *Heat wave: A social autopsy of disaster in Chicago*. University of Chicago Press.

Klinenberg, E. (2015). *Heat wave: A social autopsy of disaster in Chicago* (2nd ed.). University of Chicago Press.

Klinenberg, E. (2018). *Palaces for the people: How social infrastructure can help fight inequality, polarization, and the decline of civic life*. Crown.

Kolner, C., & Engbersen, G. (2024). Vitale sociale infrastructuur bereikt burgers en redt levens. *Sociale Vraagstukken*. https://www.socialevraagstukken.nl/vitale-sociale-infrastructuur-bereikt-burgers-en-redt-levens/

Korsten, A. F. A., & de Goede, P. (2006). De betekenis van sociaal kapitaal voor vertrouwen in overheidsbestuur. In A. F. A. Korsten & P. de Goede (Eds.), *Bouwen aan vertrouwen in het openbaar bestuur* (pp. 175–191). Elsevier.

Meijer, F. (2022). *Verbonden door rampspoed: Rampen en natievorming in negentiende-eeuws Nederland*. Uitgeverij Verloren.

Meyer, M. A. (2018). Social capital in disaster research. In H. Rodríguez, W. Donner, & J. E. Trainor (Eds.), *Handbook of Disaster Research* (pp. 263–286). Springer International Publishing. https://doi.org/10.1007/978-3-319-63254-4_14

Miles-Novelo, A., & Anderson, C. A. (2019). Climate change and psychology: Effects of rapid global warming on violence and aggression. *Current Climate Change Reports, 5*(1), 36–46. https://doi.org/10.1007/s40641-019-00121-2

NOS. (2021, July 15). *Kabinet noemt overstromingen Zuid-Limburg formeel een ramp, overheid betaalt deel schade*. Nederlandse Omroep Stichting. https://nos.nl/collectie/13869/artikel/2389502-kabinet-noemt-overstromingen-zuid-limburg-formeel-een-ramp-overheid-betaalt-deel-schade

Ntontis, E., Drury, J., Amlôt, R., Rubin, G. J., Williams, R., & Saavedra, P. (2020). Collective resilience in the disaster recovery period: Emergent social identity and observed social support are associated with collective efficacy, well-being, and the provision of social support. *British Journal of Social Psychology*, *60*(3), 1075–1095. https://doi.org/10.1111/bjso.12434

Oostveen, M. (2023, juni 8). Samen sterk? Burgercollectieven lossen problemen op die de overheid laat liggen. *de Volkskrant*. https://www.volkskrant.nl/kijkverder/v/2023/samen-sterk-burgercollectieven-lossen-problemen-op-die-de-overheid-laat-liggen~v751222/

Ostrom, E., & Ahn, T. -K. (2009). The meaning of social capital and its link to collective action. *Handbook of social capital: The troika of sociology, political science and economics*, 17–35.

Park, R. J. (2024). *Slow Burn: The Hidden Costs of a Warming World*. Princeton University Press.

Platform duurzaam Zwolle. (n.d.). *Organisaties*. https://platformduurzaamzwolle.nl/organisaties

Provincie Noord-Brabant. (2018). *65 groene daken in de wijk Kattenbosch in Rosmalen*. https://www.klimaatadaptatiebrabant.nl/voorbeelden/voorbeelden-detail/208/65-groene-daken-in-de-wijk-Kattenbosch-in-Rosmalen

Putnam, R. D. (1993). Making democracy work: Civic traditions in modern Italy. *Princeton University Press*. https://doi.org/10.2307/j.ctt7s8r7

Putnam, R. D. (2000). *Bowling alone: The collapse and revival of American community*. Simon & Schuster.

Rackin, H. M., & Weil, F. (2015). *Social capital and the repopulation of New Orleans after Hurricane Katrina*. Annual Meeting of the Population Association of America, San Diego. https://paa2015.populationassociation.org/papers/152366

RIVM. (2024). *Gezondheidseffecten van klimaatverandering: Actualisatie van de huidige klimaatrisico's voor gezondheid*. Rijksinstituut voor Volksgezondheid en Milieu. https://www.rivm.nl/publicaties/gezondheidseffecten-van-klimaat-verandering-actualisatie-van-huidige-klimaatrisicos-voor-gezondheid

Sanz-Barbero, B., Linares, C., Vives-Cases, C., González, J. L., López-Ossorio, J. J., & Díaz, J. (2018). Heat wave and the risk of intimate partner violence. *Science of The Total Environment*, *644*, 413–419. https://doi.org/10.1016/j.scitotenv.2018.06.368

Schmeets, H., & Exel, J. (2022). *De Heitjes en het vertrouwen in instituties in Heerlen*. Centraal Bureau voor de Statistiek. https://www.cbs.nl/nl-nl/longread/aanvullende-statistische-diensten/2022/de-heitjes-en-het-vertrouwen-in-instituties-in-heerlen/2-vertrouwen-als-bouwsteen-van-sociale-cohesie

Schmeets, H., Parigini, M., & van Hoof, M. (2016). *Het sociaal kapitaal in Europese regio's*. Centraal Bureau voor de Statistiek. https://www.cbs.nl/nl-nl/achtergrond/2016/43/het-sociaal-kapitaal-in-europese-regio-s

SCP. (2023). *Samenleving in beweging*. Sociaal en Cultureel Planbureau. https://www.scp.nl/publicaties/publicaties/2023/08/07/samenleving-in-beweging

SCP. (2024a). *Factsheet Hoe staat het ervoor met de sociale cohesie in Nederland?* Sociaal en Cultureel Planbureau. https://www.scp.nl/publicaties/publicaties/2024/06/04/hoe-staat-ervoor-met-de-sociale-cohesie-in-nederland

SCP. (2024b). *De leefwerelden van arm en rijk*. Sociaal en Cultureel Planbureau. https://www.scp.nl/publicaties/publicaties/2024/04/05/de-leefwerelden-van-arm-en-rijk

SCP. (n.d.). *Verkenning van toekomstige ontwikkeling van sociale cohesie*. Sociaal en Cultureel Planbureau. https://www.scp.nl/onderzoeksprogramma/verkenning-van-toekomstige-ontwikkeling-van-sociale-cohesie

Slijper, T., Urquhart, J., Poortvliet, P. M., Soriano, B., & Meuwissen, M. P. M. (2022). Exploring how social capital and learning are related to the resilience of Dutch arable farmers. *Agricultural Systems*, *198*, 103385. https://doi.org/10.1016/j.agsy.2022.103385

Smith, J. W., Anderson, D. H., & Moore, R. L. (2012). Social capital, place meanings, and perceived resilience to climate change. *Rural Sociology*, *77*(3), 380–407. https://doi.org/10.1111/j.1549-0831.2012.00082.x

Snel, E., el Farisi, B., Engbersen, G., & Krouwel, A. (2022). Sociaaleconomische status en institutioneel vertrouwen in een tijd van corona. *Tijdschrift Sociologie*, *3*.

Snel, E., Engbersen, G., Boom, G. E. J. de, & Seidler, Y. (2024). *Nederland na de pandemie: Over de doorwerking van de coronapandemie op sociale cohesie.* https://www.risbo.nl/media/119293

Snel, E., Engbersen, G., de Boom, J., Seidler, Y., Sam, N., Enneking, G., l'Istelle, F., van Heck, L., & van Linder, J. (2023). *Corona & Cohesie: Dashboard Sociale Impact Corona. Verslag meting 5 (juni 2022) en meting 6 (november 2022).* Risbo & Erasmus School of Social and Behavioural Sciences.

Szreter, S. (2002). The State of Social Capital: Bringing Back in Power, Politics, and History. *Theory and Society, 31*(5), 573–621. http://www.jstor.org/stable/3108542

Terhoeve, M. (2023). *Ode aan de moddermeisjes.* https://www.eilandennieuws.nl/watersnoodramp/watersnoodramp/14465/ode-aan-de-moddermeisjes

Tiemeijer, W. (2017). *Wat is er mis met maatschappelijke scheidslijnen?* Wetenschappelijke Raad voor het Regeringsbeleid. Verkenning nr.35. https://www.wrr.nl/documenten/2017/01/12/wat-is-er-mis-met-maatschappelijke-scheidslijnen---35

Townshend, I., Awosoga, O., Kulig, J., & Fan, H. (2014). Social cohesion and resilience across communities that have experienced a disaster. *Natural Hazards*, 76(2), 913–938. https://doi.org/10.1007/s11069-014-1526-4

Universiteit van Nederland. (2022). Videocollege: waarom zijn mensen in de zomer agressiever? *GGZ Nieuws.* https://www.ggznieuws.nl/videocollege-waarom-zijn-mensen-in-de-zomer-agressiever/

Van den Berg, K., Kolen, B., & Rudolph, M. (2022). *Evacuatiegedrag van getrof- fenen tijdens de overstromingen in Limburg in juli 2021* (PR3591.60). HKV Lijn in water. https://open.rijkswaterstaat.nl/open-overheid/onderzoeksrapporten/@257015/evacuatiegedrag-getroffenen-tijdens/

Van Holm, E. J., & Wyczalkowski, C. K. (2018). Gentrification in the wake of a hurricane: New Orleans after Katrina. *Urban Studies, 56*(13), 2763–2778. https://doi.org/10.1177/0042098018800445

Van Houwelingen, P. (2016). Indicatoren voor sociale cohesie: Het speciale geval van Japan en zijn implicaties. *Mens & Maatschappij, 91*(2), 153–171.

Van Maanen, L., Van der Mijn, R., Van Beurden, M. H. P. H., Roijendijk, L. M. M., Kingma, B. R. M., Mileti, S., & Van Rijn, H. (2019). Core body temperature speeds up temporal processing and choice behavior under deadlines. *Scientific Reports, 9*(1), 10053. https://doi.org/10.1038/s41598-019-46073-3

Verbond van Verzekeraars. (2022). *Daan Prevoo, burgemeester van Valkenburg.* https://www.verzekeraars.nl/publicaties/longreads/daan-prevoo-burgemeester-van-valkenburg

Williams, M. N., Hill, S. R., & Spicer, J. (2015). The relationship between temperature and assault in New Zealand. *Climatic Change, 132*(4), 559–573. https://doi.org/10.1007/s10584-015-1438-7

WRR. (2018). *De nieuwe verscheidenheid. Toenemende diversiteit naar herkomst in Nederland* (verkenningen nr.38). Wetenschappelijke Raad voor het Regeringsbeleid. https://www.wrr.nl/publicaties/verkenningen/2018/05/29/de-nieuwe-verscheidenheid

Zhang, F., De Dear, R., & Hancock, P. (2019). Effects of moderate thermal environments on cognitive performance: A multidisciplinary review. *Applied Energy, 236*, 760–777. https://doi.org/10.1016/j.apenergy.2018.12.005

Chapter 5
The Role of Socio-Economic Position in Climate Events

Contents

Key Points

- Socio-economically vulnerable groups are more vulnerable to many climate threats.
- Lack of resources is a barrier to taking measures that can prevent climate-related damage and to carrying out recovery efforts.
- Climate events can deteriorate socio-economic positions further, for example in the housing market and in education.

R. El-Dardiry et al., *People and Climate*, Research for Policy,
https://doi.org/10.1007/978-3-032-25831-1_5

5.1 Everyone Feels the Heat, but Some Feel it more than Others

Daisy sighs and pants. Despite keeping her balcony doors open for many hours, her apartment in the city of Nijmegen simply will not cool down. She has already endured a hot summer, with temperatures sometimes exceeding 30 °C and nights that brought little relief.[1] Daisy is not the only one. Dutch homes are often designed to let in as much sunlight as possible, but that same sun can cause temperatures to rise sharply during a heatwave.[2] Especially when airing is not an option, the heat lingers. This is also the case in Daisy's apartment. Her windows can only be tilted open by 10 cm—a small gap that offers little relief.

During this period, Daisy suffers from heat stress. This term refers to 'health complaints that arise when people are no longer able to regulate their body temperature properly due to high temperatures and humidity'.[3] Heat stress is a serious public health issue and leads to excess mortality in the summer.[4] However, the impact of heat stress extends much further. People who are unable to sleep well for several nights experience difficulties in many areas. These include increased tensions within families, poorer academic performance at school or reduced productivity. Because of climate change, this problem is expected to become much more significant in the coming decades.

Research by the investigative journalism platform Investico has shown that there are no major differences between Dutch income groups in terms of how heat-resistant their homes are.[5] Nevertheless, households with lower incomes are generally more vulnerable.[6] International research has shown that they are more likely to live in districts with less green space. As a result, these areas cool down less quickly during hot periods and the risk of heat stress is greater.[7] In addition, they are more likely to live in rental properties,[8] making them dependent on the owner for any modifications to the home. Although heat stress is a major concern for social housing tenants, a large proportion of housing associations currently have no policy in place to address heat-related issues.[9] Homeowners with limited financial means often do not have the funds to make adjustments to their own homes, let alone escape the heat.[10] Lastly,

---

[1] Daisy was extensively interviewed for the Dutch radio programme *Argos*; see Van den Adel and Bekker (2023).

[2] Daisy's 'house with a through room' is used as an example here. This term refers to homes where sunlight can shine through the entire living room. See TNO (n.d.); NOS (2024b).

[3] RIVM (n.d.)

[4] RIVM (2021); Ballester et al. (2023); CBS (2023).

[5] Of households with low and above-average incomes, 53% and 57% respectively live in a home with an excessively high heat score; see Investico (2024) and Salmons and Woutersen (2024).

[6] Van Merwijk et al. (2023).

[7] Chakraborty et al. (2019); De Vries et al. (2020); Rocha et al. (2024).

[8] See e.g. SCP (2020).

[9] BZK. (2023).

[10] Van Merwijk et al. (2023).

individuals with a lower socio-economic position are often more susceptible to heat-related health problems, as they are more likely than average to suffer from obesity and cardiovascular diseases.[11]

This overview of heat-related issues demonstrates that the impacts of climate change are distributed unequally. In some scenarios, however, climate change may also reduce socio-economic problems, for example if rising temperatures result in milder winters. This would reduce the likelihood of households falling into energy poverty, as well as excess mortality.[12]

This chapter focuses on the link between the consequences of climate change and socio-economic position. We pay particular attention to socio-economically vulnerable groups. There is often a link between this vulnerability and other forms of social inequality, for example in the areas of health and participation.[13] First, we explain how scientists analyse socio-economic position (Sect. 5.2). We then examine how socio-economic position helps determine the impact of climate events (Sect. 5.3). Next, we consider risks that increase socio-economic disparities in society, focusing among other things on the delayed effects of climate events (Sect. 5.4). Illustration 5.1 shows that socio-economic position can influence the impact of a climate event, such as a heat wave.

5.2 Socio-Economic Position and Vulnerabilities

Socio-economic differences between neighbourhoods exist in almost all Dutch municipalities. The Hague offers a good example. Traditionally, the more affluent have lived on higher sandy soils, while people with a lower socio-economic position have lived on low-lying peat soils.[14] The location of one's home is also a status symbol. Scientists and policymakers have long been concerned with the question of how to address socio-economic differences in society.[15]

In order to study the socio-economic circumstances of households, scientists have been looking for indicators that can express socio-economic position. In this context, the term socio-economic status score (SES score) is often used.[16] Statistics Netherlands uses data on prosperity, education and recent employment history for this purpose. Together, these data provide an overview of differences in social positions within society. Of course, such a picture is not complete; other factors are also

[11] RIVM (2012: 44); Schwartz et al. (2020).

[12] For an example in which the impact of a severe winter is recently calculated, see CPB (2022). For an analysis of excess mortality, see RIVM (2023); KNMI (2023).

[13] RIVM (2012); De Voogd and Cuperus (2021).

[14] The Hague Historical Museum (n.d.); Municipality of The Hague (n.d. b).

[15] WRR (2017a).

[16] For definitions of socio-economic status, see VZinfo (n.d.).

Illustration 5.1 Socio-economic position

important, such as political participation or—as the example of The Hague illustrates—the postcode of a residence.[17] In this chapter, we therefore primarily view SES scores as measures that can be used to *approximate* socio-economic positions and differences.[18]

When we look at socio-economic position in this way, we see considerable differences across the Netherlands (Fig. 5.1). In particular, the average socio-economic position of households is low in the north-east of the country, the far south-east and the major cities (shown on the left). The major cities also show greater inequality (shown on the right). The aforementioned divide in The Hague, for example, is clearly visible when looking at socio-economic status scores at the neighbourhood level.[19] Rotterdam, Amsterdam and The Hague are the municipalities where the most people live in poverty in both relative and absolute terms (between 10 and 11% of the local population in 2017).[20]

Socio-economic position is often statistically associated with other forms of social vulnerability. For example, a study by the National Institute for Public Health and the Environment shows that people with a lower SES score, on average, suffered more from the effects of the coronavirus pandemic. They were overrepresented in excess mortality figures.[21]

This greater vulnerability was already present at the outset: on average, people with a low SES score have poorer health, smoke more often and are more often overweight.[22] Similar correlations in disaster studies make clear that other variables often play an important role, such as age, ethnicity, family situation, gender, language proficiency or the location of the home.[23] The accumulation of such variables increases vulnerability.[24]

Statistical association does not imply that individual households with a similar SES score exhibit the same degree of vulnerability. For example, a wealthy household with a villa in a floodplain is more susceptible to potential flooding than a low-income household living in a high-rise apartment.[25]

---

[17] CBS (2021a).

[18] We note here that prosperity concerns both household income and assets.

[19] Gemeente Den Haag (n.d. a).

[20] SCP (2019).

[21] RIVM (2023: 49).

[22] Springvloet et al. (2017); Schwartz et al. (2020).

[23] Flanagan et al. (2011).

[24] Westerhof et al. (2023).

[25] An example of spatial vulnerabilities that are not related to socio-economic status can be found in Arnhem; see Forrest et al. (2020).

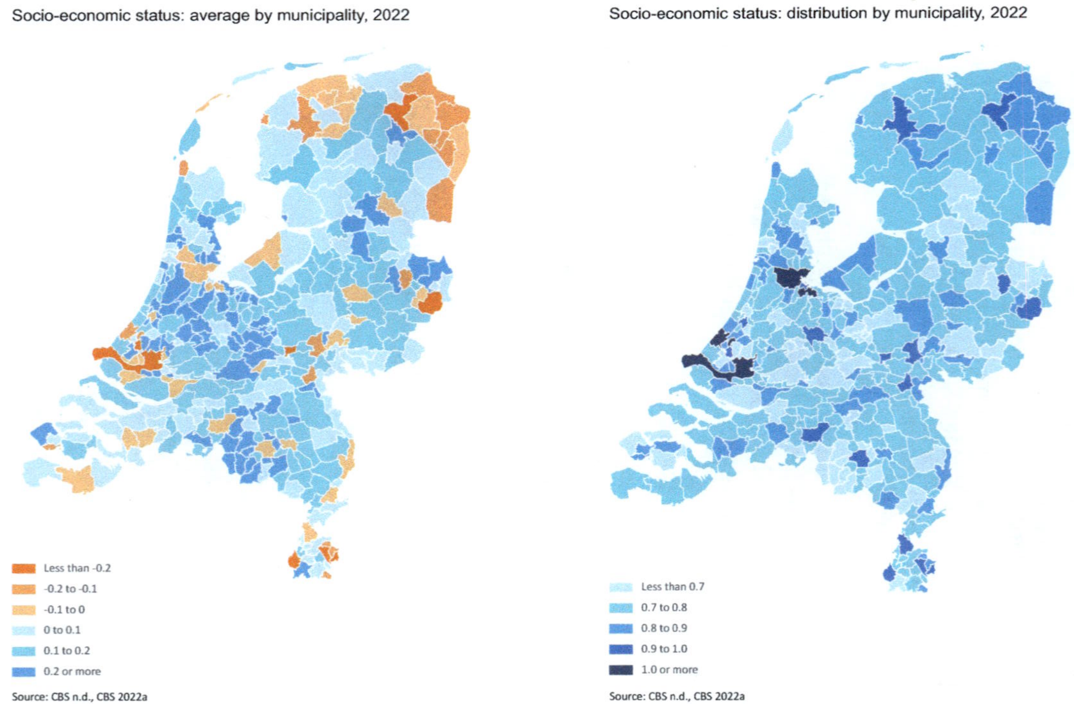

Fig. 5.1 Socio-economic position: average and distribution by municipality.[26] *Source* CBS (n.d.), CBS (2022)

[26] CBS (2022). The average socio-economic status score for the Netherlands as a whole is 0. For a detailed discussion of what these scores mean and how they are determined, see CBS (n.d.)

5.3 The Impact of a Vulnerable Socio-Economic Position on Preparation, Response and Recovery

In this section, we examine how climate vulnerability is related to a vulnerable socio-economic position during climate events. Naturally, it is not the case that poorer population groups within a given area experience different weather than groups living a few kilometres away. However, the circumstances in which people with a more vulnerable socio-economic position live mean that, overall, they are more likely to experience negative impacts from climate change.[27] A climate event is therefore not only a physical phenomenon, but also a social event with a range of social and societal consequences.[28] The events during and after hurricane Katrina in the American city of New Orleans in 2005 illustrate this point (see Box 5.1).

Box 5.1 The impact of hurricane Katrina

An example of a natural disaster that disproportionately affected vulnerable communities is hurricane Katrina in 2005. This hurricane of the most severe category caused extensive damage in New Orleans and the surrounding area. Although the link to climate change was not directly emphasised at the time, the disaster serves as a warning for future storms.[29] The hurricane caused widespread damage throughout the city, but it was the low-lying districts with many low-income residents and minority groups that were hit the hardest.[30] These were precisely the districts with a relatively high proportion of residents with low incomes, many of whom were African Americans. Public recovery efforts in these districts also progressed more slowly than in the wealthier, predominantly white districts.[31]

Social cohesion—the subject of the previous chapter—was also a determining factor in the impact. For example, a severely affected Vietnamese district recovered significantly faster than neighbouring, wealthier districts. The church, where much of community life took place, played a crucial role. By maintaining contact with various members of the community, the church was able to coordinate how and when residents returned. This reduced the risk that people would end up in an abandoned district without services or neighbours. At a later stage, members of the community joined forces through the church to stand united against political plans that were hindering the recovery of the district.[32]

---

[27] Westerhof et al. (2023).

[28] Tavery and Wagner-Pacifici (2022); Quarantelli (1986: 2).

[29] Lynn et al. (2009); Cannavò (2008).

[30] Frank (2020); Bolin and Kurz (2018).

[31] Islam and Winkel (2017).

[32] Chamlee-Wright and Store (2009).

Box 5.1 (continued)

These examples of the unequal impact and resolution of damage and recovery demonstrate that climate vulnerability is not only determined by physical conditions (such as the location of the home, the severity of storms and rainfall), but also by social factors, including socio-economic position and social cohesion.[33]

The greater impact of climate change on socio-economically vulnerable groups manifests itself in four ways: through (1) increased exposure and sensitivity; (2) limited adaptive capacity; (3) fewer perspectives for action during climate events; and (4) lower recovery capacity.[34] We will discuss these four aspects in turn.

5.3.1 Increased Exposure and Sensitivity

Exposure to climate threats is primarily determined by where someone lives: for example, people in low-lying areas are more exposed to flood risk, while those in densely built-up areas without green spaces are more exposed to heat stress. Social inequality is often reflected in the living environment. This is also evident in the Netherlands. Districts where households have above-average incomes are more likely to be greener, for example because homes have gardens.[35]

More trees and plants and fewer paved surfaces offer at least two advantages: rainwater is better absorbed into the soil, reducing the risk of flooding, and more greenery mitigates heat. Various Dutch studies show that ‘heat islands’ can be up to 9 °C warmer than their surroundings during hot periods.[36] Such heat islands can be disastrous. Regardless of any measures people may take themselves, those living in greener environments are less exposed to the negative effects of heat and extreme rainfall. In addition to the physical environment, the type of work that people do can also increase exposure to climate events.[37] People who work indoors are less likely to be affected by heat, cold or rain. By contrast, workers in physically demanding sectors such as construction, agriculture or emergency services are more exposed to hazards during extreme weather.[38] Moreover, their productivity decreases in such situations, with negative consequences for the income of groups like the

---

[33] Klinenberg (2002).

[34] This classification is based on various studies: Islam and Winkel (2017); PBL (2024a); Forrest et al. (2020). Other classifications are also possible, for example from disaster preparedness literature: prevention, preparation, response and recovery; see Bonfanti et al. (2024).

[35] De Vries et al. (2020); Maas et al. (2009).

[36] TNO (2012); RIVM (2012); Atlas of the Living Environment (2020); Urban Heat Islands (n.d.).

[37] Islam and Winkel (2017).

[38] Park (2024).

self-employed.[39] A survey of around 500 arable farmers shows that the majority are experiencing declining incomes due to increasing weather extremes. Despite measures such as drainage and irrigation, their financial vulnerability is increasing.[40]

Exposure alone does not necessarily result in climate-related damage. Whether a climate event leads to damage also depends on sensitivity. Increased sensitivity among socio-economically vulnerable groups is often the result of a combination of factors. For example, there is a clear link between socio-economic position and health. People with lower levels of education are more likely to smoke and to be overweight.[41] In other words, they are more likely to belong to epidemiological risk groups.[42] These health disparities in turn influence the impact of climate events on individuals. For example, people who are overweight have more difficulty dissipating heat and are therefore more susceptible to heat stress.[43] In addition to socio-economic position, other factors also play a role in sensitivity, such as gender, age, physical disabilities and ethnicity.[44]

Finally, there is often an accumulation of problems. People living in poverty are not only more vulnerable to climate risks, but also frequently face a multitude of other challenges.[45] These problems require time, attention and resources.[46] Climate change may be recognised as a problem, but it often ranks low on the list of priorities, as a resident of the Caribbean island of Bonaire illustrates:

> I can't worry about climate change. That's a problem for the future. Right now I have to feed my children.[47]

A cumulative burden of challenges reduces people's resilience to adversity and increases their vulnerability to climate threats.

5.3.2 Limited Adaptive Capacity

Climate vulnerability is not a given. Good preparation—which reduces exposure or sensitivity—can prevent much suffering. For example, people can adapt the layout of their homes to minimise damage in the event of flooding. Seemingly simple measures can already make a difference, such as placing sandbags at the door and avoiding laminate flooring on the ground floor. However, some preparations, such

[39] Hallegatte and Rozenberg (2017).

[40] Hiddink (2023); NOS (2024a).

[41] CBS (2025); CPB (2013); WRR (2018).

[42] De Vries et al. (2020); Ministry of the Interior and Kingdom Relations et al. (2023); Chakraborty et al. (2019).

[43] Speakman (2018).

[44] Van Merwijk et al. (2023); Levac et al. (2012).

[45] Westerhof et al. (2023).

[46] Westerhof et al. (2023).

[47] Caribbean Network (2021). Translated by the WRR.

as installing sunshades or air conditioning, involve high costs. Other measures do not require money, but do demand risk awareness and knowledge. Examples include seeking reliable information, considering evacuation options and escape routes and responding promptly to warning signals.

Research into the taking of precautionary measures against potential flooding in the city of Arnhem illustrates how social inequality affects adaptive capacity. In the Burgemeesterswijk—a district with high rates of home ownership and households with a high socio-economic position—resilience is generally high. In the Spijkerkwartier and the city centre, where incomes are lower and there are many rental properties, the situation is different.

In these districts, there is less enthusiasm for taking measures. The researchers linked this to the high percentage of rental properties and the fear of having to reverse adaptations or incurring possible fines when moving.[48]

Socio-economic position is an important predictor of whether or not people take preparatory and adaptation measures.[49] This applies both to measures that require financial investment and to those that do not.[50] Broadly speaking, there are two main barriers to preparation: lack of knowledge and lack of resources.

Knowledge begins with awareness. People who are aware of a threat are more likely to seek information about it and possibly take preparatory measures. The risk and crisis barometer from spring 2023 shows that the number of people gathering information about events such as extreme weather and flooding is increasing. A quarter of people sought information about extreme weather.[51] The risk perception of climate threats among citizens varies according to socio-economic position. According to research by Statistics Netherlands, in 2020, 85% of highly educated people saw climate change as a major problem for the future, compared to 67% among those with lower levels of education.[52] Other statistical research concerns adaptation preferences among residents in Rotterdam. It shows, among other things, that people with lower levels of education are significantly less concerned about the risk of flooding and heat.[53] An international study on flood risk found that older people and those with lower levels of education are less likely to seek relevant information on websites.[54] Without awareness and the right knowledge, people are unlikely to take action to protect themselves against a changing climate. However, even greater knowledge does not guarantee that more adequate preparatory measures will be taken (see also Chap. 3).[55]

In addition to information about threats, specific knowledge about regulations and subsidy applications for adaptation measures also plays a role. Not everyone is

[48] Forrest et al., (2020: 8).

[49] Levac et al. (2012); Hallegatte and Rozenberg (2017).

[50] Zamboni and Martin (2020).

[51] I&O Research (2023).

[52] I&O Research (2023).

[53] CBS (2021b).

[54] Derkzen et al. (2017).

[55] O'Sullivan et al. (2012).

aware of the available options. Moreover, applications can be complex. For example, residents who apply for a subsidy to disconnect rainwater downpipes from the sewer may be required to provide information about their property, such as site plans or photographs.[56] Whether people apply for such government subsidies is also related to socio-economic position. In the city of Breda, for example, residents have been able since April 2020 to apply to the municipality for partial reimbursement of the costs of climate adaptation measures. These include greening gardens, installing sunshades and disconnecting rainwater downpipes. On average, subsidy applicants have higher incomes and higher levels of education.[57] A study on the provision of subsidies in the municipality of Arnhem has also shown that it is mainly affluent residents who benefit.[58] A similar mechanism is at play in the allocation of subsidies for the energy transition, where wealthier households in particular receive subsidies that are partially funded by less well-off households. Some of the subsidised measures also make homes more climate-resilient, specifically in the case of insulation.[59]

A lack of financial resources is a second barrier to implementing climate adaptation measures. Even when subsidies are available, there is still an amount that must be paid by citizens or businesses themselves. Not everyone has the means to do so. Furthermore, residents of rental properties are unable to make changes to their homes themselves. As noted at the beginning of this chapter, housing associations and other landlords are often still insufficiently interested in climate adaptation. To put it bluntly, socio-economically vulnerable groups are more likely to be dependent on others and therefore experience a reduced capacity to act.[60]

5.3.3 Fewer Perspectives for Action During Climate Events

In earlier centuries, wealthy Amsterdam merchants would leave the city during the summer months. They built country estates for this purpose along the Vecht and Amstel rivers and in the dunes near the North Sea coast, each more impressive and larger than the last.[61] While their poorer fellow citizens had to endure the stench rising from the filthy canals and were exposed to infectious diseases such as plague and cholera, the wealthy enjoyed fresh air in the countryside. Those with means had the option to escape the annual malaise.

The stinking canals may seem a thing of the distant past, but even in recent times there are examples of situations in which people's socio-economic position limits their perspectives for action, with potentially negative consequences for their well-being. Take the energy crisis, for example. Rising gas and electricity prices have led

[56] WRR (2017b).

[57] Van den Broek et al. (2020).

[58] Van Valkengoed et al., (2020; 2021).

[59] Gerretschen (2022).

[60] WRR (2023).

[61] Canon of the Netherlands (n.d.); Municipality of Amsterdam (2024a).

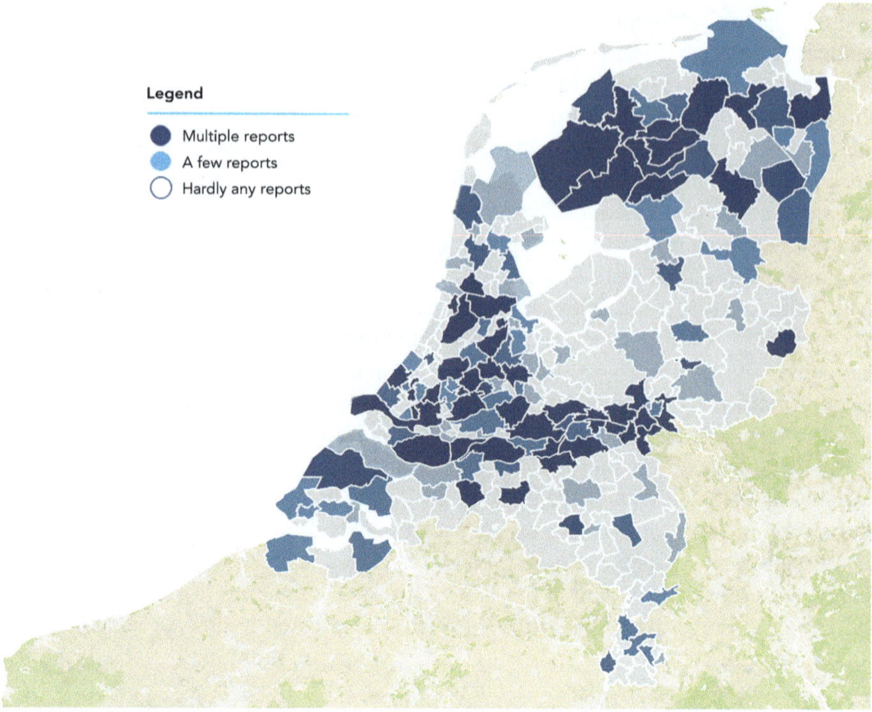

Fig. 5.2 Spatial distribution of reports of foundation problems in the Netherlands.[64] *Source* KCAF n.d. a

to increasing energy poverty, particularly in the areas shown in Fig. 5.2 as having lower socio-economic status scores.[62] When energy costs become unaffordable for a household, it is not surprising that households turn the thermostat down significantly or even switch it off. Although it is difficult to demonstrate a causal relationship, Dutch researchers have found that health care costs are higher among energy-poor households that consume little energy.[63]

Fear of high bills can literally be fatal. A notorious example of the vulnerability of households with limited financial means dates from 1980. A severe heatwave struck the Midwest of the United States. Local authorities distributed free fans to citizens in need, mostly elderly people and low-income households. But because these households feared high energy bills, the fans were not used.[65] During the heatwave, around 150 people died. Most were poor and elderly.

[62] TNO (2023a). Energy poverty is defined as 'when a household has a low income in combination with a high energy bill or a dwelling of (very) poor energy quality', see Netherlands Organisation for Applied Scientific Research (2023a).

[63] TNO (2023b).

[64] KCAF (n.d. a).

[65] US House of Representatives Testimony (1980: 36).

Access to information is also crucial during crisis situations. However, not all information reaches every target group equally. For example, people with limited proficiency in Dutch or with hearing or visual impairments may have difficulty understanding evacuation details.[66]

Another concern regarding perspectives for action during a crisis is the level of trust citizens have in government. This trust is linked to socio-economic position.[67] Citizens with less trust in the authorities are less likely to follow advice.[68] As a consequence, their vulnerability increases.

If a disaster is imminent or a climate event occurs, evacuation may be necessary. People will then be asked to leave their homes. During flooding in the town of Valkenburg in the summer of 2021, between 1,500 and 2,000 people were evacuated.[69] Residents without access to a car found it more difficult to evacuate safely, especially if they lacked a social network to rely on.[70] In general, people with a more vulnerable socio-economic position are less able to evacuate, even if they have received the appropriate information.[71] Wealthier citizens have more options to relocate physically or to stockpile emergency supplies. Car ownership enables people to evacuate themselves and their loved ones more quickly. With sufficient financial resources, it is also easier to arrange temporary accommodation elsewhere.[72]

5.3.4 Lower Recovery Capacity

Inequalities also persist after a disaster. Even with equal exposure and vulnerability, the extent to which someone is able to recover can determine how 'unequal' people emerge from a crisis situation.[73] Here, we address both physical and mental recovery.

Repairing damaged homes and buildings is costly. Funding may come from personal savings and private insurance or from government or disaster relief funds.[74] It is evident that households with substantial financial reserves are able to start repairs or replace household goods more quickly.[75] If a household has limited financial

66 Flanagan et al. (2011).

67 SCP (2023a); Fornalé et al. (2023); Bolin and Kurz (2018).

68 Bonfanti et al. (2024); Stel et al. (2017); Fothergill and Peek (2004).

69 Van Mersbergen (2021).

70 Van den Berg et al. (2022)

71 Fothergill and Peek (2004).

72 Gladwin and Peacock (1997).

73 Islam and Winkel (2017).

74 In the event of a disaster, such as a flood, the Dutch National Disaster Relief Fund may initiate a fundraising campaign. Private individuals can contribute to this effort. The aim is to support those most severely affected. An example is the Limburg Flood Recovery Fund; see Dutch National Disaster Relief Fund (n.d.).

75 Leichenko and Silva (2014); Comerio et al. (1994).

means, recovery is more likely to be postponed.[76] Socio-economic differences also play a role in insurance coverage. Less affluent households are less likely to be insured against damage caused by natural disasters.[77] Those in less favourable circumstances are also less likely to access recovery fund budgets, as the application process is often complex and time-consuming (see also the example on subsidy applications in Sect. 5.3).[78]

The aftermath of the flooding in the province of Limburg painfully illustrates how socio-economic inequalities affect recovery. A resident of Geulle, one of the flooded villages, described it as follows:

> There are people here on the street who really had to wait until the insurance paid out before they could start [renovating]. It simply wasn't possible. So you're almost a year and a half further down the line, because everything takes so long.[79]

On top of the differences in the speed of recovery they face because of their financial circumstances, some people can find recovery a major psychological challenge. International research has shown that trust in the authorities reduces stress during emergencies and that strong community cohesion often leads to less post-traumatic stress following a disaster (see also Chap. 4).[80] If victims feel abandoned during a disaster, this can have a lasting and devastating impact on their trust in the authorities involved. People with low levels of trust tend to have poorer health and fewer social contacts.[81] Here too, people in vulnerable social positions generally have less trust in both government and others.[82] As a result, they are at greater risk of slower psychological recovery.

Many evacuated citizens were still struggling with psychological problems a year after the flood in the province of Limburg. This was partly related to the slow pace of repairs and the resulting lack of time and mental and physical space to process the events. The loss of control and sense of security, as well as the disappearance of everyday routines, can weigh heavily on some victims.[83]

---

[76] Islam and Winkel (2017).

[77] Thomas et al. (2018); Botzen (2017); Cannon et al. (2020).

[78] Fothergill and Peek (2004).

[79] Van den Berg et al. (2022: 34). Translated by the WRR.

[80] Bonfanti et al. (2024); Matthews et al. (2020).

[81] Thoresen et al. (2018).

[82] SCP (2023a).

[83] Van den Berg et al. (2022).

5.4 Effects of Climate Change that Further Worsen a Vulnerable Socio-Economic Position

So far in this chapter, we have examined how socio-economic position influences where and how the impacts of climate change are felt within a society. We have mainly looked at climate events that occur rapidly, such as a flood.[84] However, climate change can result not only in rapid and sudden events, but also in slow, almost creeping changes.[85] For example, models show that the shift in climate zones over the next 50 years will be greater than in the past 6,000 years.[86] The shifting of climate zones also regularly makes the news. The hot French summers of the 1970s are now quite normal in the Netherlands.[87] Such changes have consequences for a wide range of sectors, from tourism and agriculture to health, sports and education.[88] This change is taking place gradually. The likelihood of warm summers is increasing, while the chance of another Elfstedentocht—the traditional skating tour along 11 cities in the province of Friesland—is diminishing.[89]

In a similarly insidious way, climate events can contribute to increasing socio-economic vulnerability. In this section, we explain two mechanisms that can deteriorate the socio-economic position of vulnerable groups further. This does not mean that every climate event necessarily leads to greater socio-economic disparities. As noted at the beginning of this chapter, higher winter temperatures can reduce the risk of energy poverty for socio-economically vulnerable groups. In addition, we discuss two case studies. We show that some groups in the housing market and in education may be more vulnerable as a result of climate change.

5.4.1 Mechanisms that can Deteriorate a Vulnerable Socio-Economic Position Further

Because socio-economically vulnerable groups experience greater exposure and have more limited adaptive and response capacity, there is a risk that their social position will deteriorate further after a climate event. For example, they may incur debt to repair their homes, leading to new financial problems. Wealthier individuals are able to 'bounce back' more quickly and effectively after climate events. Some researchers refer to this as a vicious cycle, as described in a United Nations study:

> It begins with multidimensional inequalities, which then cause greater exposure of the vulnerable groups to climate hazards, *increase their susceptibility to* damage caused by

[84] Erikson (1995).

[85] Erikson (1995); Boin et al. (2021).

[86] Xu et al. (2020).

[87] NRC (2022).

[88] KNMI (2023).

[89] Visser et al. (2025).

> these hazards, and decrease their ability to cope with and recover from the damage. As a result, when the climate hazards actually hit, vulnerable groups suffer disproportionate loss of income and assets (physical, financial, human, and social). Climate change thus makes inequality worse, thus perpetuating the cycle.[90]

The consequences of climate change can undermine the socio-economic security of vulnerable groups further through rising prices.[91] Furthermore, climate-related events such as prolonged drought and excessive rainfall are becoming more frequent, increasing the risk of smaller or failed harvests.[92] Various studies indicate that climate change will impact global food production, particularly crop yields; producer incomes; and the prices, availability, quality and safety of food.[93] For example, the price of olive oil rose by several tens of percentage points in 2023 and 2024 due to persistent drought in the Mediterranean region.[94] Households with limited financial resources are more severely affected by rising food prices, as they spend a relatively larger share of their income on food.[95]

Many households in the Netherlands also have limited margins to absorb price increases. We need only look back a short time to understand the effects that price increases in essential goods can have. In 2022 and 2023, the rise in energy prices posed a significant risk for poorer population groups. The Netherlands Bureau for Economic Policy Analysis (CPB) has repeatedly pointed out that, without additional policy measures, large groups of Dutch citizens are at risk of falling below the poverty line.[96]

Households with a strong socio-economic position are generally more resilient, while those with a more vulnerable position are, on average, more vulnerable to climate impacts. Persistent vulnerability threatens to erode the economic position of poorer households. Chain effects represent a significant risk in this regard, as we demonstrate using the following case studies of climate-vulnerable districts and education.

5.4.2 *Case 1: The Consequences of Climate Change for the Housing Market*

Climate threats are not evenly distributed across the country, as we have argued several times in this chapter. For many homeowners—private owners and housing associations—drought is an insidious problem that can cause significant financial damage in the long term. In many parts of the Netherlands, building foundations are

[90] Islam and Winkel (2017: 7).

[91] Carleton and Hsiang (2016).

[92] Schlenker and Roberts (2009).

[93] Vermeulen et al. (2012); Carleton and Hsiang (2016).

[94] Eurostat (2024).

[95] Leichenko and Silva (2014).

[96] CPB (2023).

vulnerable, for example because low groundwater levels have caused the wooden foundation piles to rot. In addition, some homes are at risk of damage from subsidence, due in part to the type of ground on which they are built. The predicted increase in summer drought will exacerbate foundation problems.[97] Exposure to this risk is not evenly distributed in the Netherlands, as reflected in regional differences in damage reports. Figure 5.2 illustrates this.

The Council for the Environment and Infrastructure (Rli) estimates that by 2035, around 425,000 buildings in the Netherlands will be affected by foundation problems.[98] After that, the expected numbers increase, but so does the uncertainty. During dry periods, the Knowledge Centre for Foundation Problems (KCAF) receives a dozen to several dozen serious reports of foundation damage from across the Netherlands every day.[99] The problem already exists, and climate change will increase the number of future reports. Foundation damage is a major financial setback for owners. Repairs can cost between €50,000 and €100,000.[100]

Such an amount is easier to afford for households with greater financial means than for those with lower incomes or less wealth. According to the Housing Cost Monitor of the National Mortgage Guarantee, one in five homeowners with a mortgage had difficulty paying their mortgage costs.[101,102] The repair costs associated with foundation damage pose a significant risk for this group.[103] Foundation damage is an example of how climate change can exacerbate inequalities on the housing market. However, the impact on the housing market is broader and more far-reaching, as the development of 'climate gentrification' in the US clearly demonstrates. There, homes in higher-lying districts near the coast are more expensive than those in low-lying ones.[104] This higher valuation stems from the lower risk of flooding. The higher property values have led to a different composition of districts and neighbourhoods. Wealthier households choose to live at higher elevations, while poorer households settle in the lower-lying districts.

This creates a negative feedback loop in which a vulnerable socio-economic position and climate vulnerability reinforce each other. Such an effect is not yet visible in the Netherlands, but climate-vulnerable districts already exist. These districts are more exposed to climate risks than average, and many of their residents have a

---

[97] Rli (2024).

[98] Rli (2024: 27).

[99] KCAF (n.d.a).

[100] KCAF (n.d. b.); ABN AMRO (2023).

[101] The National Mortgage Guarantee is a guarantee that provides homebuyers with security. It serves as a safety net for mortgagors who are unable to meet their payments due to certain circumstances.

[102] National Mortgage Guarantee (2024). Moreover, risks are exacerbated by the tight housing market. Buyers are bidding above asking prices and taking out maximum mortgages. To increase their chances of a successful offer, many buyers forgo a structural survey; see NVM (2024: 3). This situation can result in buyers purchasing a home without financial reserves and facing high repair costs after a few years.

[103] NVM (2024).

[104] Keenan et al. (2018); De Koning and Filatova (2020).

low socio-economic position.[105] As climate risks become increasingly apparent, it is likely that these risks will be factored into housing prices. Households that can afford it will move to climate-resilient districts, while homes in climate-vulnerable districts will become financially accessible to households with fewer resources. This mechanism increases the socio-economic disparities between climate-resilient and climate-vulnerable districts.[106]

5.4.3 Case 2: The Consequences of Climate Change for Education

The inequality-amplifying effects of climate change can also manifest in education. Heat is particularly problematic. A warm classroom significantly reduces academic performance.[107] A school year that is half a degree warmer can result in a 1% decrease in learning outcomes.[108] The negative effect of heatwaves is even greater.[109] In the United States, the impact of weather on academic performance is observed across all groups, but the negative effect is about three times greater for African American and Latino students. This is related to the fact that these students are more likely to live in districts with a pronounced urban heat island effect.[110]

The Dutch Inspectorate of Education has also conducted research in the Netherlands on the impact of heat on academic performance. Scores for final exams in secondary education are 0.034 points lower on hot days compared to days with normal temperatures. This may seem minor, but these are average effects. According to the Inspectorate, the magnitude of this effect is 'slightly more than half the size of the average effect of well-studied interventions for mathematics or reading in secondary education".[111] In current government policy, however, climate adaptation in schools receives little attention (see Box 5.2).

---

[105] ABN AMRO (2023).

[106] De Koning and Filatova (2020: 2); European Economic Area (2024).

[107] Ebi et al. (2021).

[108] Wargocki et al. (2019); Park et al. (2020).

[109] Park et al. (2020).

[110] Park et al., (2020: 310).

[111] Inspectorate of Education (2022a: 53, 2022b).

Box 5.2 Adaptation measures in Dutch schools

Although climate change is addressed in education policy, adaptation measures to reduce vulnerability to extreme weather are barely on the agenda. One example is the *Implementation Agenda for Climate & Energy*.[112] Much attention is devoted to making school buildings more sustainable, for example through insulation and the installation of solar panels. However, adaptation measures in schools are not mentioned. Terms such as 'cooling', 'air conditioning' and 'sunshade' do not appear in the implementation agenda.

At the provincial and local levels, there is some attention to adaptation. For example, there are websites in various provinces that inform schools about the climate-resilient redesign of schoolyards, such as in the provinces of South Holland, North Brabant and Limburg.[113] Each initiative promotes 'green–blue' measures—in other words, removing paving and creating more space for plants and water storage. The provinces also make budgets available to implement these redesigns. There are also various specific projects at the municipal level. In some municipalities, for example, partnerships are being set up between design agencies, architects and the municipality itself to create climate-adaptive schools.

As shown by international comparative research, lower exam results affect educational outcomes in the long term. This effect is more pronounced among young children and students living in poor districts.[114] A likely reason for the lagging performance of these groups is that they have little or no access to resources that mitigate heat, such as air conditioning at home.[115] A cooler home or a cooler school can make a significant difference in terms of educational opportunities. In other words, inequality of opportunity, combined with heat, plays a significant role early in life.

5.5 Conclusion

We began this chapter with the example of Daisy. In her rental apartment, it is difficult to take measures that would make hot summers more bearable. The situation is quite different in the eastern part of her hometown Nijmegen, in the Kwakkenberg

---

[112] OCW (2023).

[113] Green–Blue Schoolyards (n.d.); Green–Blue Revolution Schoolyards (n.d.); Schoolyard of the Future (n.d.).

[114] Prentice et al. (2024).

[115] Park et al. (2020: 20). The authors refer to 'less access to potentially compensatory resources'. However, they do not specify what these resources are.

district—a leafy neighbourhood with mainly detached houses and a wooded environment. People own large gardens in which they can seek out the shade and children can splash and play happily in swimming pools. Heat stress is hardly an issue here. Wealthier homeowners are more likely to have sufficient means to adapt their homes: they can install underfloor cooling, purchase air conditioning or fit their windows with sunshades.

How people experience climate change is partly determined by their socio-economic position. In this chapter, we have discussed examples showing that the consequences of climate change hit socio-economically vulnerable groups particularly hard. The starting position of households with a vulnerable socio-economic position is often worse, and their ability to make adjustments is limited. When climate events occur, factors such as level of education, type of work, income and assets all influence the impact and the speed of recovery. This creates mechanisms that deteriorate socio-economic positions further and ultimately widen socio-economic disparities.

If people experience financial hardship due to a climate event, underlying discontent can develop into the previously described feelings of resentment. The distrust that often accompanies this resentment offers little prospect of coping well with the next climate event, especially when people withdraw into themselves. In short, climate resilience requires continuous attention to the resilience of society. All social factors play a role in this.

References

ABN AMRO. (2023). *Stapeling klimaatrisico's en financiële draagkracht op de woningmarkt*. https://www.abnamro.com/research/nl/onze-research/stapeling-klimaatrisicos-en-financiele-draagkracht-op-de-woningmarkt

Atlas Leefomgeving. (2020). *Stedelijk hitte-eiland effect (UHI)*. https://www.atlasleefomgeving.nl/stedelijk-hitte-eiland-effect-uhi

Ballester, J., Quijal-Zamorano, M., Méndez Turrubiates, R. F., Pegenaute, F., Herrmann, F. R., Robine, J. M., Basagaña, X., Tonne, C., Antó, J. M., & Achebak, H. (2023). Heat-related mortality in Europe during the summer of 2022. *Nature Medicine, 29*(7), 1857–1866. https://doi.org/10.1038/s41591-023-02419-z

Boin, A., Ekengren, M., & Rhinard, M. (2021). *Understanding the creeping crisis*. Springer Nature.

Bolin, B., & Kurtz, L. C. (2018). Race, Class, Ethnicity, and Disaster Vulnerability. In H. Rodríguez, W. Donner, & J. E. Trainor (Eds.), *Handbook of Disaster Research* (pp. 181–203). Springer International Publishing. https://doi.org/10.1007/978-3-319-63254-4_10

Bonfanti, R. C., Oberti, B., Ravazzoli, E., Rinaldi, A., Ruggieri, S., & Schimmenti, A. (2024). The Role of Trust in Disaster Risk Reduction: A Critical Review. *International Journal of Environmental Research and Public Health, 21*(1). https://doi.org/10.3390/ijerph21010029

Botzen, W. J. W. (2017). *Economie van klimaatverandering en natuurrampen*. VU.

BZK. (2023). *Hittestress: Er is meer aandacht, maar een eenduidige aanpak ontbreekt nog*. Ministerie van Binnenlandse Zaken en Koninkrijksrelaties. https://www.klimaatweb.nl/wp-content/uploads/po-assets/876118.pdf

Cannavò, P. F. (2008). In the Wake of Katrina: Climate Change and the Coming Crisis of Displacement. *Political Theory and Global Climate Change*, 177–200. http://ndl.ethernet.edu.et/bitstream/123456789/29172/1/8.pdf.pdf#page=204

Cannon, C., Fox Gotham, K., Lauve-Moon, K., & Powers, B. (2020). The climate change double whammy: Flood damage and the determinants of flood insurance coverage, the case of post-Katrina New Orleans. *Climate Risk Management, 27*, Article 100210. https://doi.org/10.1016/j.crm.2019.100210

Canon of the Netherlands / Canon van Nederland. (n.d.). *De stinkende stad ontvluchten.* https://www.canonvannederland.nl/nl/noord-holland/onderwijscanon-amstelland-en-de-meerlanden/de-stin-107984706

Caribisch Netwerk. (2021). *Klimaatverandering voor Bonairiaan ver van bed show: 'armoede overheerst'.* https://caribischnetwerk.ntr.nl/2021/10/25/klimaatverandering-voor-bonairiaan-ver-van-bed-show-armoe-de-overheerst/

Carleton, T. A., & Hsiang, S. M. (2016). Social and economic impacts of climate. *Science, 353*(6304), aad9837.

CBS. (2021a). *Berekenwijze Sociaal Economische Status scores.* Centraal Bureau voor de Statistiek. https://www.cbs.nl/nl-nl/maatwerk/2021/45/berekenwijze-ses-score-per-wijk-buurt

CBS. (2021b). *Klimaatverandering en energietransitie: opvattingen en gedrag van Nederlanders in 2020.* Centraal Bureau voor de Statistiek. https://www.cbs.nl/nl-nl/longread/rapportages/2021/klimaatverandering-en-energietransitie-opvattingen-en-gedrag-van-nederlanders-in-2020?onepage=true#c-2—Opvattingen-over-klimaatverandering

CBS. (2022). *Statusscore per wijk en buurt o.b.v. welvaart, opleidingsniveau en arbeid.* Centraal Bureau voor de Statistiek. https://www.cbs.nl/nl-nl/achtergrond/2022/14/statusscore-per-wijk-en-buurt-o-b-v-welvaart-opleidingsniveau-en-arbeid

CBS. (2023). *Oversterfte in juni.* Centraal Bureau voor de Statistiek. https://www.cbs.nl/nl-nl/nieuws/2023/28/oversterfte-in-juni

CBS. (2025). *Leefstijl en sociaaleconomische status.* Centraal Bureau voor de Statistiek. https://www.cbs.nl/nl-nl/longread/statistische-trends/2025/leefstijl-en-sociaaleconomische-status

CBS. (n.d.). *Hoe interpreteer je de SES-WOA-scores en hoe zijn deze bepaald*? Centraal Bureau voor de Statistiek. https://www.cbs.nl/nl-nl/faq/infoservice/hoe-interpreteer-je-de-ses-woa-scores-en-hoe-zijn-deze-bepaald-

Chakraborty, T., Hsu, A., Manya, D., & Sheriff, G. (2019). Disproportionately higher exposure to urban heat in lower-income neighborhoods: a multi-city perspective. *Environmental Research Letters*, 14, 105003. https://doi.org/10.1088/1748-9326/ab3b99/meta

Chamlee-Wright, E., & Storr, V. H. (2009). Club goods and post-disaster com-munity return. *Rationality and Society, 21*(4), 429–458. https://ssrn.com/abstract=1711318

Comerio, M. C., Landis, J. D., & Rofe, Y. (1994). *Post-disaster residential rebuilding: a study for California governor's Office of Emergency Services (OES).* (Working paper, Vol. 608). University of California.

CPB. (2013). *Toekomst voor de zorg.* Centraal Planbureau. https://www.cpb.nl/publicatie/toekomst-voor-de-zorg

CPB. (2022). *Ontwikkeling van het arbeidsinkomen tijdens corona.* Centraal Planbureau. https://www.cpb.nl/sites/default/files/omnidownload/CPB-Coronapublicatie-Ontwikkeling-van-het-arbeidsinkomen-tijdens-corona.pdf

CPB. (2023). *Concept Macro Economische Verkenning 2024.* Centraal Planbureau. https://www.cpb.nl/sites/default/files/omnidownload/CPB-Raming-Concept-Macro-Economische-Verkenning-2024-augustus-2023.pdf

Dutch National Disaster Relief Found/Nationaal Rampenfonds. (n.d.). *Het Nationaal Rampenfonds.* https://nationaalrampenfonds.nl/

De Koning, K., & Filatova, T. (2020). Repetitive floods intensify outmigration and climate gentrification in coastal cities. *Environmental Research Letters, 15*(3), Article 034008.

De Voogd, J. & Cuperus, R. (2021). *Atlas van afgehaakt Nederland. Over buitenstaanderse en gevestigden.*

De Vries, S., Buijs, A. E., & Snep, R. P. H. (2020). Environmental Justice in The Netherlands: Presence and Quality of Greenspace Differ by Socioeconomic Status of Neighbourhoods. *Sustainability, 12*(15), 5889. https://www.mdpi.com/2071-1050/12/15/5889

Derkzen, M. L., Van Teeffelen, A. J. A., & Verburg, P. H. (2017). Green infrastructure for urban climate adaptation: How do residents' views on climate impacts and green infrastructure shape adaptation preferences? *Landscape and Urban Planning, 157*, 106–130. https://doi.org/10.1016/j.landurbplan.2016.05.027

Ebi, K. L., Vanos, J., Baldwin, J. W., Bell, J. E., Hondula, D. M., Errett, N. A., Hayes, K., Reid, C. E., Saha, S., Spector, J., & Berry, P. (2021). Extreme Weather and Climate Change: Population Health and Health System Implications. *Annual Review of Public Health, 42*, 293–315. https://doi.org/10.1146/annurev-publhealth-012420-105026

EEA. (2024). *European Climate Risk Assessment*. European Environment Agency. https://www.eea.europa.eu/en/analysis/publications/european-climate-risk-assessment#:~:text=The%20first%20European%20Climate%20Risk%20Assessment%20%28EUCRA%29%20identifies,infrastructure%2C%20water%20resources%2C%20financial%20stability%2C%20and%20people%E2%80%99s%20health

Erikson, K. (1995). *A new species of trouble: The human experience of modern disasters*. Norton & Company Inc.

Eurostat. (2024). *Price of olive oil up 50% in one year*. https://ec.europa.eu/eurostat/web/products-eurostat-news/w/ddn-20240227-1

Flanagan, B. E., Gregory, E. W., Hallisey, E. J., Heitgerd, J. L., & Lewis, B. (2011). A social vulnerability index for disaster management. *Journal of Homeland Security and Emergency Management, 8*(1), 0000102202154773551792.

Forrest, S. A., Trell, E. -M., & Woltjer, J. (2020). Socio-spatial inequalities in flood resilience: Rainfall flooding in the city of Arnhem. *Cities, 105*, 102843. https://doi.org/10.1016/j.cities.2020.102843

Fothergill, A., & Peek, L. A. (2004). Poverty and disasters in the United States: A review of recent sociological findings. *Natural Hazards, 32*(1), 89–110. https://doi.org/10.1023/B:NHAZ.0000026792.76181.d9

Frank, T. (2020). Flooding disproportionately harms black neighborhoods. *Scientific American*. https://www.scientificamerican.com/article/flooding-disproportionately-harms-black-neighborhoods/

Gemeente Den Haag. (n.d. a). *Den Haag in cijfers*. https://denhaag.incijfers.nl/home?ReturnUrl=%2fdashboard%2fnl-nl%2fsegregatiemonitor%2f3—wonen-en-leefomgeving

Gemeente Den Haag. (n.d. b). *Leeftijd, achtergrond en sociaal-economische status*. https://denhaag.incijfers.nl/dashboard/nl-nl/segregatiemonitor/2—leeftijd—achtergrond-en-sociaal-economische-status

Gerretschen, J. (2022). Klimaatadaptatie in het vaarwater van sociale gelijkheid. *Water Governance Tijdschrift*. https://www.stowa.nl/sites/default/files/assets/PROJECTEN/Projecten%202022/Water%20Governance/Edities/WAGO_2022-02_Klimaatactie_DEF.pdf

Gladwin, H., & Peacock, W. G. (1997). Warning and evacuation: A night for hard houses. In W. G. Peacock, & H. Gladwin (Eds.), *Hurricane Andrew. Ethnitic y, Gender and the Sociology of Disasters* (pp. 52–74). Routledge.

Green–Blue Schoolyards/Groenblauwe Revolutieschoolpleinen. (n.d.). *Schoolpleinen*. https://groenblauwerevolutieschoolpleinen.nl

Green–Blue Revolution Schoolyards/Groenblauwe Schoolpleinen. (n.d.). *Samen maken we onze schoolpleinen duurzaam*. https://www.groenblauweschoolpleinen.nl

Hallegatte, S., & Rozenberg, J. (2017). Climate change through a poverty lens. *Nature Climate Change, 7*(4), 250–256. https://doi.org/10.1038/nclimate3253

Hiddink, J. (2023). *Ruim helft van akkerbouwers kampt met opbrengstderving door weersextremen*. Nieuwe Oogst. https://www.nieuweoogst.nl/nieuws/2024/05/03/ruim-helft-van-akkerbouwers-kampt-met-opbrengstder-ving-door-weersextremen

I&O Research (2023). *Risico- en crisisbarometer voorjaar 2023* (2023/092). Nationaal Coördinator Terrorismebestrijding en Veiligheid (NCTV). https://www.nctv.nl/documenten/rapporten/2023/06/05/risico--en-crisisbarometer---voorjaar-2023

Inspectorate of Education/Inspectie van het Onderwijs. (2022a). *De Staat van het Onderwijs 2022*. https://www.onderwijsinspectie.nl/documenten/rapporten/2022/04/13/de-staat-van-het-onderwijs-2022

Inspectorate of Education/Inspectie van het Onderwijs. (2022b). *Technisch rapport huisvesting*. https://www.onderwijsinspectie.nl/documenten/rapporten/2022/04/13/technisch-rapport-huisvesting-svho-2022

Investico. (2024). *Onderzoeksverantwoording Hittestress*. https://www.platform-investico.nl/media/pages/onderzoeken/2-miljoen-ouderen-in-woning-met-risico-op-oververhitting/4b2cd50847-1720512693/onderzoeksverantwoording-hittestress.pdf

Islam, N. W., & Winkel, J. (2017). *Climate change and social inequality* (DESA Working Paper, No. 152). Verenigde Naties, Department of Economic & Social Affairs. https://bhekisisa.org/wp-content/uploads/2023/09/wp152_2017.pdf

KCAF. (n.d. a). *Landelijk loket funderingsproblematiek*. Kennis Centrum Aanpak Funderingsproblematiek. https://www.kcaf.nl/landelijk-loket-funderingsproblematiek/

KCAF. (n.d. b). *De Nationale Funderingsramp (deel 2): Funderingsherstel vraagt een lange adem*. Kennis Centrum Aanpak Funderingsproblematiek. https://www.kcaf.nl/de-nationale-funderingsramp-deel-2/

Keenan, J. M., Hill, T., & Gumber, A. (2018). Climate gentrification: From theory to empiricism in Miami-Dade County, Florida. *Environmental Research Letters, 13*(5), Article 054001.

Klinenberg, E. (2002). *Heat wave: A social autopsy of disaster in Chicago*. University of Chicago Press.

KNMI. (2023). *KNMI'23-klimaatscenario's*. Koninklijk Nederlands Meteorologisch Instituut. https://www.knmi.nl/kennis-en-datacentrum/achtergrond/knmi-23-klimaatscenario-s

Leichenko, R., & Silva, J. A. (2014). Climate change and poverty: vulnerability, impacts, and alleviation strategies. *WIREs Climate Change, 5*(4), 539–556. https://doi.org/10.1002/wcc.287

Levac, J., Toal-Sullivan, D., & O'Sullivan, T. L. (2012). Household Emergency Preparedness: A Literature Review. *Journal of Community Health, 37*(3), 725–733. https://doi.org/10.1007/s10900-011-9488-x

Lynn, B. H., Healy, R., & Druyan, L. M. (2009). Investigation of Hurricane Katrina characteristics for future, warmer climates. *Climate Research, 39*(1), 75–86.

Maas, J., Verheij, R. A., de Vries, S., Spreeuwenberg, P., Schellevis, F. G., & Groenewegen, P. P. (2009). Morbidity is related to a green living environment. *Journal of Epidemiology and Community Health, 63*(12), 967–973. https://doi.org/10.1136/jech.2008.079038

Matthews, V., Longman, J., Bennett-Levy, J., Braddon, M., Passey, M., Bailie, R. S., & Berry, H. L. (2020). Belonging and inclusivity make a resilient future for all: A cross-sectional analysis of post-flood social capital in a diverse Australian rural community. *International Journal of Environmental Research and Public Health, 17*(20), 7676.

Municipality of Amsterdam/Gemeente Amsterdam. (2024). *Op de vlucht voor poepvliegen en cholera: de Amsterdamse buitenplaatsen van weleer*. https://www.amsterdam.nl/nieuws/achtergrond/buitenplaatsen/

National Mortgage Guarantee / Nationale Hypotheek Garantie (NHG). (2024). *De Woonlastenmonitor*. https://www.nhg.nl/woonlastenmonitor/

NIPV. (2019). *Maatschappelijk herstel na grootschalige overstromingen*. Nationaal Instituut Publieke Veiligheid. https://nipv.nl/wp-content/uploads/2022/02/20191017-IFV-Wave2020-Maatschappelijk-herstel-na-grootschalige-overstromingen.pdf

NOS. (2024a, May 3). *Bijna twee derde akkerbouwers zegt minder te verdienen door weersextremen*. Nederlandse Omroep Stichting. https://nos.nl/artikel/2519067-bijna-twee-derde-akkerbouwers-zegt-minder-te-verdienen-door-weersextremen

NOS. (2024b, August 12). *Niet kunnen slapen door de hitte: 'Koelte-ongelijkheid steeds groter probleem'*. Nederlandse Omroep Stichting. https://nos.nl/artikel/2533041-niet-kunnen-slapen-door-de-hitte-koelte-ongelijkheid-steeds-groter-probleem

NRC. (2022, July 22). *Klimaatverschuiving: die lange hete Franse zomers uit de jaren 70 vind je nu in Nederland*. https://www.nrc.nl/nieuws/2022/07/22/klimaatverschuiving-die-lange-hete-franse-zomers-uit-de-jaren-70-vind-je-nu-in-nederland-2-a4137119

NVM. (2024). *Analyse woningmarkt 2e kwartaal 2024*. Nederlandse Vereniging van Makelaars. https://www.nvm.nl/media/lf5f1152/bijlage-1-analyse-nvm-woningmarkt-2e-kwartaal-2024.pdf

O'Sullivan, J. J., Bradford, R. A., Bonaiuto, M., De Dominicis, S., Rotko, P., Aaltonen, J., Waylen, K., & Langan, S. J. (2012). Enhancing flood resilience through improved risk communications. *Natural Hazards and Earth System Sciences, 12*(7), 2271–2282.

OCW. (2023). *Uitvoeringsagenda Klimaat & Energie*. Ministerie van Onderwijs, Cultuur en Wetenschap. https://www.rijksoverheid.nl/documenten/rapporten/2023/10/13/uitvoeringsagenda-klimaat-energie

Park, R. J. (2024). *Slow Burn: The Hidden Costs of a Warming World*. Princeton University Press.

Park, R. J., Goodman, J., Hurwitz, M., & Smith, J. (2020). Heat and learning. *American Economic Journal: Economic Policy, 12*(2), 306–339.

Prentice, C. M., Vergunst, F., Minor, K., & Berry, H. L. (2024). Education outcomes in the era of global climate change. *Nature Climate Change, 14*(3), 214–224.

Quarantelli, E. L. (1986). *Planning and management for the prevention and mitigation of natural disasters, especially in a metropolitan context: Initial questions and issues which need to be addressed*. https://udspace.udel.edu/server/api/core/bitstreams/6bd84606-41a0-43a9-a49b-f21a7b678e6a/content

Rijksmuseum. (n.d.). *IJsvermaak bij een stad*. http://hdl.handle.net/10934/RM0001.COLLECT.495302

RIVM. (2012). GGD-*richtlijn medische milieukunde: Gezondheidsrisico's van zomerse omstandigheden*. Rijksinstituut voor Volksgezondheid en Milieu. https://www.rivm.nl/bibliotheek/rapporten/609400007.pdf

RIVM. (2021). *Klimaatverandering leidt nu al tot meer sterfte door hitte*. Rijksinstituut voor Volksgezondheid en Milieu. https://www.rivm.nl/nieuws/klimaatverandering-leidt-nu-al-tot-meer-sterfte-door-hitte

RIVM. (2023). *Gezondheidseconomische aspecten van de covidpandemie*. Rijksinstituut voor Volksgezondheid en Milieu. https://www.rivm.nl/bibliotheek/rapporten/2023-0318.pdf

RIVM. (n.d.). GGD-*richtlijn medische milieukunde: hitte en gezondheid*. Rijksinstituut voor Volksgezondheid en Milieu. https://www.rivm.nl/ggd-richtlijn-mmk-hitte-gezondheid

Rli. (2024). *Goed gefundeerd*. Raad voor de leefomgeving en infrastructuur. https://www.rli.nl/publicaties/2024/advies/goed-gefundeerd

Rocha, A. D., Vulova, S., Förster, M., Gioli, B., Matthews, B., Helfter, C., Meier, F., Steeneveld, G., Barlow, J., Järvi, L., Chrysoulakis, N., Nicolini, G., & Kleinschmit, B. (2024). Unprivileged groups are less served by green cooling services in major European urban areas. *Nature Cities, 1*(2024), 424–435.

Salmons, M., & Woutersen, E. (2024, juli 10). De doorzonwoning als een hitte-eiland: Goed geïsoleerd, maar bloedjeheet. *De Groene Amsterdammer*. https://www.groene.nl/artikel/goed-geisoleerd-maar-bloedjeheet

Schlenker, W., & Roberts, M. J. (2009). Nonlinear temperature effects indicate severe damages to U.S. crop yields under climate change. *Proceedings of the National Academy of Sciences, 106*(37), 15594–15598. https://doi.org/10.1073/pnas.0906865106

Schoolyard of the Future/Schoolplein van de toekomst. (n.d.). *Schoolplein van de toekomst*. https://schoolpleinvandetoekomst.nl

Schwaller, N. L., Kelmenson, S., BenDor, T. K., & Spurlock, D. (2020). From abstract futures to concrete experiences: How does political ideology interact with threat perception to affect climate adaptation decisions? *Environmental Science & Policy, 112*, 440–452.

Schwartz, T., Luiten, W., Kok, L. (2020). *Gezondheidsproblemen personen met een lage sociaal-economische status. SEO Economisch Onderzoek.* https://www.seo.nl/wp-content/uploads/2021/01/2021-07-Gezondheidsproblemen-personen-met-een-lage-sociaal-economische-status.pdf

SCP. (2019). *Waar wonen de armen in Nederland?* Sociaal en Cultureel Planbureau. https://digitaal.scp.nl/armoedeinkaart2019/waar-wonen-de-armen-in-nederland/

SCP. (2020). *De sociale staat van Nederland Wonen.* https://digitaal.scp.nl/ssn2020/wonen/

SCP. (2023a). *Eigentijdse ongelijkheid: De postindustriële klassenstructuur op basis van vier typen kapitaal.* Sociaal en Cultureel Planbureau. https://www.scp.nl/publicaties/publicaties/2023/03/07/eigentijdse-ongelijkheid

SCP. (2023b). *Samenleving in beweging.* Sociaal en Cultureel Planbureau. https://www.scp.nl/publicaties/publicaties/2023/08/07/samenleving-in-beweging

Speakman, J. R. (2018). Obesity and thermoregulation. *Handbook. Clinical Neurology, 156*, 431–443. https://doi.org/10.1016/b978-0-444-63912-7.00026-6

Springvloet, L., Kuipers, M., & Van Laar, M. (2017). *Effecten van stoppen-met- roken ondersteuning onder rokers met een lage sociaal economische status.* https://www.trimbos.nl/aanbod/webwinkel/af1559-effecten-van-stoppen-met-roken-ondersteuning-onder-rokers-met-een-lage-sociaal-economische-status/

Stel, M., Ketelaar, D., Gutteling, J., Giebels, E., & Kerstholt, J. (2017). *Het alarmeren en informeren van kwetsbare groepen bij crisissituaties.* Universiteit Twente, Vakgroep Psychologie van Conflict, Risico en Veiligheid.

Tavory, I., & Wagner-Pacifici, R. (2022). Climate change as an event. *Poetics, 93*, Article 101600. https://doi.org/10.1016/j.poetic.2021.101600

The Hague Historical Museum /Haags Historisch Museum. (n.d.). *Stad van zand en veen.* https://www.haagshistorischmuseum.nl/nl/ontdek/verhalen-over-den-haag/zand-en-veen

Thomas, K., Hardy, R. D., Lazrus, H., Mendez, M., Orlove, B., Rivera-Collazo, I., Roberts, J. T., Rockman, M., Warner, B. P., & Winthrop, R. (2018). Explaining differential vulnerability to climate change: A social science review. *Wires Climate Change, 10*(2), Article e565. https://doi.org/10.1002/wcc.565

Thoresen, S., Birkeland, M., Wentzel-Larsen, T., & Blix, I. (2018). Loss of Trust May Never Heal. Institutional Trust in Disaster Victims in a Long-Term Perspective: Associations With Social Support and Mental Health. *Frontiers in Psychology, 9*, 1–10. https://doi.org/10.3389/fpsyg.2018.01204

TNO. (2023a). *Energiearmoede in Nederland 2022.* https://www.rijksoverheid.nl/documenten/rapporten/2023/01/27/tno-rapport-energiearmoede-in-nederland-2022

TNO. (2023b). *Gezondheidskosten en energiearmoede: Een empirische analyse voor Nederland.* https://publications.tno.nl/publication/34640440/YyUGjI/TNO-2023-gezondheidsrisico.pdf

TNO. (n.d.). *Factsheet Climate Proof Cities: Hittestress.* https://www.tno.nl/media/3959/factsheet-hittestress.pdf

Urban Heat Islands (n.d.). Waar zijn ze? https://hitte-eilanden.nl/waar-zijn-ze/

U.S. House of Representatives Testimony. (1980). *Federal Efforts to Aid Low-income and Elderly Individuals Affected by Life-Threatening Heat Conditions.* U.S. Government Printing Office.

Van den Adel, I., & Bekker, B. (2023). *Daisy heeft last van hittestress.* Argos. https://www.vpro.nl/argos/lees/onderwerpen/thermo-staat/2023/daisy-heeft-last-van-hittestress.html

Van den Berg, K., Kolen, B., & Rudolph, M. (2022). *Evacuatiegedrag van getrof-fenen tijdens de overstromingen in Limburg in juli 2021* (PR3591.60). HKV Lijn in water. https://open.rijkswaterstaat.nl/open-overheid/onderzoeksrapporten/@257015/evacuatiegedrag-getroffenen-tijdens/

Van den Broek, M., J. Harms., F. Smit. (2020). *Gedragsaanpak: Financiële prikkels voor klimaatadaptieve maatregelen.* Behavioural Insights Nederland/Inspire to Act. https://klimaatadaptatienederland.nl/@240732/gedragsaanpak-financiele-prikkels/

Van Mersbergen, S. (2021). Half miljard euro schade in Limburg, opruimen kan wel vier maanden duren. https://www.ad.nl/binnenland/half-miljard-euro-schade-in-limburg-opruimen-kan-wel-vier-maanden-duren~a1db2afa/

Van Merwijk, K. (2023). *Verkoeling bij hitte* Academische Werkplaats Gezonde Leefomgeving. https://awgl.nl/projecten/verkoeling-bij-hitte

Van Valkengoed, A., Kreuger, M., Geerts, B., & Heessels, M. (2021). *Vervolgonderzoek Subsidie voor klimaatadaptatie.* Rijksuniversiteit Groningen.

Van Valkengoed, A., Kreuger, M., Peters, A., Geerts, B., & Heessels, M. (2020). *Subsidies voor klimaatadaptatie.* Gemeente Breda.

Vermeulen, S. J., Campbell, B. M., & Ingram, J. S. I. (2012). Climate Change and Food Systems. *Annual Review of Environment and Resources, 37*, 195–222. https://doi.org/10.1146/annurev-environ-020411-130608

Visser, H., Van Prooije, H., De Vries, H., & Petersen, A.C. (2025). The likelihood of holding outdoor skating marathons: the past, present and future of a climate-change indicator, and a way to adapt. *Climate Change, 178.* https://doi.org/10.1007/s10584-025-03920-9

VZinfo. (n.d.). *Wat is sociaaleconomische status?* https://www.vzinfo.nl/sociaaleconomische-status/verantwoording/definities

Wargocki, P., Porras-Salazar, J. A., & Contreras-Espinoza, S. (2019). The relationship between classroom temperature and children's performance in school. *Building and Environment, 157*, 197–204. https://doi.org/10.1016/j.buildenv.2019.04.046

Westerhof , S., Taanman, M., Van Popering-Verkerk, J., & Duijn, M. (2023). *Why Inequality matters to Resilience.* Erasmus Universiteit Rotterdam. https://convergence.nl/app/uploads/RDI_veerkracht-en-ongelijkheid_V3.pdf

WRR. (2017a). *Wat is er mis met maatschappelijke scheidslijnen?* (verkenningen nr.35). Wetenschappelijke Raad voor het Regeringsbeleid. https://www.wrr.nl/publicaties/verkenningen/2017/01/12/wat-is-er-mis-met-maatschappelijke-scheidslijnen---35

WRR. (2017b). *Weten is nog geen doen. Een realistisch perspectief op redzaamheid* (rapport nr.97). Wetenschappelijke Raad voor het Regeringsbeleid. https://www.wrr.nl/publicaties/rapporten/2017/04/24/weten-is-nog-geen-doen

WRR. (2018). *Van verschil naar potentieel. Een realistisch perspectief op de sociaaleconomische gezondheidsverschillen* (policy brief). Wetenschappelijke Raad voor het Regeringsbeleid. https://www.wrr.nl/publicaties/policy-briefs/2018/08/27/van-verschil-naar-potentieel.-een-realistisch-perspectief-op-de-sociaaleconomische-gezondheidsverschillen

WRR. (2023). *Grip: Het maatschappelijk belang van persoonlijke controle* (rapport nr.108). Wetenschappelijke Raad voor het Regeringsbeleid. https://www.wrr.nl/publicaties/rapporten/2023/11/30/grip

Xu, C., Kohler, T. A., Lenton, T. M., Svenning, J.-C., & Scheffer, M. (2020). Future of the human climate niche. *Proceedings of the National Academy of Sciences, 117*(21), 11350–11355. https://doi.org/10.1073/pnas.1910114117

Zamboni, L. M., & Martin, E. G. (2020). Association of US Households' Disaster Preparedness With Socioeconomic Characteristics, Composition, and Region. *JAMA Network Open, 3*(4), Article e206881. https://doi.org/10.1001/jamanetworkopen.2020.6881

Chapter 6
Stronger Together in a Changing Climate

Contents

6.1 Towards a More Broadly Supported Adaptation Policy

Policymakers in government bodies, businesses and civil-society organisations cannot ignore the fact that the Netherlands will increasingly be exposed to the effects of both gradual changes such as sea level rise and acute threats caused by extreme weather. To understand what this may entail, we only need to look at the impact of heatwaves that struck cities and regions in France and Italy (2015), the floods in Germany, Belgium and the Dutch province of Limburg (2021) and the torrential rains in the Spanish region of Valencia (2024). Climate disasters can cause loss of life and devastation, disrupt societies for extended periods, ruin businesses and result in enormous financial and ecological damage.

The impact of climate events extends beyond physical and economic damage. Climate events also influence how people view themselves and society. In times of crisis, social cohesion can increase. This stronger cohesion makes it easier to work together on recovery and preparedness.

However, this is not a given. Climate events can also lead to a loss of trust in the authorities and the surrounding community. The coronavirus pandemic in the Netherlands has clearly shown that crisis management can unintentionally set groups of citizens against each other. In some cases, this has led individuals to lose faith

R. El-Dardiry et al., *People and Climate*, Research for Policy,
https://doi.org/10.1007/978-3-032-25831-1_6

in the democratic rule of law, turn away from government and retreat into circles of like-minded people. Similar dynamics may arise around climate policy and responses to climate events. A government that neglects social factors in adaptation policy will miss opportunities to enhance climate resilience. Moreover, if citizens feel abandoned, government policy may inadvertently contribute to social disruption. This can also undermine public support for preventive measures.

One thing is clear: investing in physical measures is and will remain essential in the coming decades. However, understanding the role of social factors in climate adaptation shows that physical measures are *necessary*, but not *sufficient* to make and keep our country climate-resilient. First, not all climate threats can be mitigated with physical measures. Some risks are inherently difficult to predict, and no measure can eliminate risks entirely. Second, social conditions and human behaviour also shape the societal impact of climate events. Successfully anticipating climate threats and managing climate events requires attention to both the physical and social context.

The more effective the government becomes at designing adaptation policy that acknowledges the role of social factors in climate events, the better the Netherlands will be able to withstand increasing climate threats. In this book, the WRR argues that *climate resilience in the Netherlands will improve if the government also invests in social infrastructure, which encourages people to look out for each other, increases mutual trust and offers perspectives for action.* Figure 6.1 provides a schematic overview of the findings and recommendations that we will discuss in more detail in this chapter.

6.2 The Importance of Social Factors

In the preceding chapters, we examined the interplay between the consequences of climate change and three social factors: dealing with uncertainty (Chap. 3), social cohesion (Chap. 4) and socio-economic position (Chap. 5). Time and again, we saw that climate events are also social events. Conversely, climate events influence social relations within affected communities. We briefly summarise our threefold analysis.

People differ in how they *deal with the uncertainty of climate threats*—the socio-psychological component. The concepts of threat appraisal and coping appraisal underpin the adoption of measures. It is unlikely that people will seek ways to protect themselves without confidence in their own abilities. What people think and do is partly shaped by their environment and by what others consider desirable or normal. Social norms, often linked to group identities, can therefore play an important role in how people deal with climate threats.

The attitudes of individuals and communities towards risks in their living environment have historical roots. Disasters from the past leave traces in the present. For instance, communities that have frequently faced floods or wildfires adapt their ways of life accordingly. On the other hand, people tend to prepare less thoroughly for threats that have not occurred in the past or that are not embedded in collective memory.

Social factors matter when it comes to climate adaptation

Dealing with uncertainty — Social cohesion — Socio-economic position

Two conclusions on the role of social factors in adaptation policy

Conclusion 1

Climate resilience has a physical and a social component, which ideally reinforce each other but are not necessarily aligned

Conclusion 2

In a climate-resilient society, people look out for each other, trust each other, and have the capacity to act

Climate resilience in the Netherlands improves if the government **also invests in social infrastructure** which encourages people to look out for each other, increases mutual trust, and offers perspectives for action.

Recommendation 1

Invest in social infrastructure to promote societal resilience

Recommendation 2

Promote the joining of forces and mutual trust

Recommendation 3

Ensure that people know what they can do to adapt to a changing climate

Fig. 6.1 Overview of findings and recommendations

Physical measures cannot eliminate all climate risks. Every physical measure, by definition, leaves residual risks. That is why climate resilience also means that a society is able to withstand disasters without excessive damage and suffering. *Social cohesion* plays a crucial role in this. Safety also arises from people being able to rely on a network and experience support within their own living environment. Neighbourhoods where people look out for each other recover more quickly after a disaster. Conversely, a climate event itself can impact social cohesion. Among other things, adversity can foster solidarity—but the effect is not necessarily positive. Social cohesion can also decline, for instance if the actions of the authorities during a climate event cause confusion or are perceived as unfair by certain groups of citizens.

Finally, groups with a weak *socio-economic position* are generally more vulnerable to climate threats. Their exposure is often relatively high, while their capacity to make adjustments to their living situation and lifestyle is relatively limited. A lack of resources forms a barrier both to taking preventive measures and to carrying out repairs after suffering damage as a result of climate events.

These three social factors are closely linked. In a manner of speaking, people who are socially isolated, live in poverty and have little knowledge of climate risks are three times as vulnerable. Climate events are highly likely to cause significant damage and suffering in the coming decades. If people have not felt adequately protected and if the costs of recovery become unaffordable, collective trauma and feelings of resentment may arise. In such a situation, groups of citizens may lose trust in the community and in government. In the worst case, a toxic cocktail of mistrust, alienation and poverty can develop.

However, this dynamic can also work the other way. Working together during disasters can confirm and even increase the collective self-confidence of communities. This is reinforced when people perceive the actions of institutions (such as government bodies, aid agencies and insurers) during and after disasters as helpful and fair.

6.3 Physical as well as Social Pathways to Climate Resilience

Dutch adaptation policy aims to accelerate the transition to 'climate resilience'.[1] This also includes the capacity to cope effectively with extreme climate events. The social factors described in this book offer a new perspective on the feasibility of this aim.

Based on its analysis of the interaction between social factors and the consequences of climate change, the WRR draws two conclusions that are relevant to adaptation policy.

[1] IenW (2023). The National Climate Adaptation Strategy will be revised in 2026.

In a Climate Resilience has a Physical and a Social Component, Which Ideally Reinforce Each Other but are not Necessarily Aligned

A country that wants to reduce its vulnerability to climate change must make progress on two fronts simultaneously: the physical and the social. A range of physical interventions provides the population with a certain level of protection. The authorities can take responsibility for some of these physical interventions, but citizens and businesses will also need to take action themselves in their own living environment and business operations. On the social front, this concerns the development of societal resilience: since complete protection against all possible climate events is an illusion, the community must also be able to cope with disasters and recover from them.

Physical protection and societal resilience therefore require efforts from the authorities, citizens, businesses and civil-society organisations. Internationally, the Netherlands is lauded as an example of how to create safe living conditions in a river delta. The long history of the Dutch water authorities demonstrates that, in addition to dykes and hydraulic engineering works, solidarity and cooperation are essential ingredients. The challenges surrounding climate adaptation represent a new chapter in safeguarding the Dutch living environment.

However, there is still much progress to be made on both the physical and the social front, partly because the necessity of climate resilience does not always carry sufficient weight among policymakers. The idea that Dutch land is literally malleable is deeply embedded in the administrative culture. This belief is evident, for example, in the approach to major challenges in housing construction and spatial planning. In policy practice, the pursuit of climate-resilient spatial planning is often overshadowed by other policy priorities, such as alleviating the housing shortage. Building sufficient housing will remain one of the main policy priorities at both national and regional levels in the coming years. As a consequence, some municipalities still choose to build part of their housing stock in front of dykes or on land where there is a real risk of subsidence.[2] Despite increasing climate threats and growing awareness, the percentage of homes in vulnerable areas in the Netherlands has risen, from 7.9% in 2000 to 8.3% in 2021 and 9.2% in 2023.[3] The official policy line in Dutch government now states that water and soil conditions must be taken into account.[4] For the time being, however, it is often not the quality of the subsoil, but the utility value of land that is decisive in land use decisions.[5]

Dutch society itself has taken limited action in the field of climate adaptation. On the whole, the Dutch are aware of climate change. They acknowledge that climate

[2] Examples include the planned districts of Stadsblokken and Meinerswijk in Arnhem and the new village of Cortelande to be built in the Zuidplaspolder (South Holland); see NRC (2024a; 2024b), respectively.

[3] Nationaal Deltaprogramma (2022); PBL (2024).

[4] 'Taking water and soil conditions into account' is a change from the previous policy principle in which water and soil conditions were leading. See IenW (2024).

[5] Rli (2024).

change is man-made and that it brings new threats. Nevertheless, taking adaptation measures has so far received relatively little priority from many citizens, civil-society organisations and businesses.

This is not without reason. After the North Sea flood of 1953, the Dutch government invested heavily and with great success in physical protection measures. Although floods have caused considerable damage since then, there have been no further fatalities. The downside of this achievement is that many citizens hardly consider the possibility that current flood protection projects might prove inadequate.[6]

Although other extremes, such as heat and drought, are relatively unknown to many Dutch people, they risk falling into a similar pattern of optimistic expectations ('the government will keep us safe, that's what we pay taxes for').Given the confidence in physical protection measures and the expectation that the government will ultimately intervene if disaster does strike, it is perhaps unsurprising that few people have heeded the call to purchase an emergency package.[7]

This makes it clear that physical measures in climate adaptation policy can also unintentionally have negative social effects. This is also known as the vulnerability paradox: because physical measures give people a strong sense of security, they no longer prepare for situations of insecurity, making them particularly vulnerable when a disaster does occur.[8] Heavy reliance on the physical measures that have been implemented can make people insensitive to the fact that all measures are accompanied by residual risks. These residual risks will increase if global warming continues to accelerate. Moreover, most physical protection measures in the Netherlands are aimed at flood protection, while physical measures against other manifestations of climate change, such as drought and heat, are still in their infancy.

Policymakers and administrators would do well to take into account the social impact that physical measures may have. In Box 6.1, we discuss several barriers and opportunities in this regard.

Box 6.1 Adaptation projects can serve multiple objectives

Large-scale infrastructure projects are the primary example of efforts to ensure physical safety. In order to provide everyone in the Netherlands behind the dykes with a basic level of protection, a major overhaul of the waterworks is underway through the Dutch Delta Programme. From 2024 to 2037, the

[6] Research shows that high confidence in sea defences and dykes influences the adoption of protective measures. Other studies indicate that few people are concerned about flooding in their immediate living environment and that many people trust the government to take sufficient protective measures. See Terpstra (2010, 2011), Van Thiel and Mol (2020), Netherlands Court of Audit (2023) https://www.nationaleombudsman.nl/system/files/bijlage/Herstel%20bieden_een%20vak%20apart.pdf and SCP (2025).

[7] AD (2022); NOS (2022); SCP (2025).

[8] Dückers (2023).

national government is investing €7.7 billion in flood protection.[9] The total budget for the already completed Room for the River programme amounted to €2.3 billion.[10] The necessity and value of such large-scale programmes and projects are substantiated by social cost–benefit analyses, in which reducing the number of casualties is given priority.[11] Efforts to prevent casualties, for example by increasing response times in high-risk areas and enabling vertical evacuation, reduce the risk of societal disruption.[12]

However, not all types of social costs and benefits are taken into account to the same extent. One reason is that some aspects of flood protection are difficult to express in monetary terms.

When initiating dyke reinforcement projects aimed at meeting flood protection standards, the principles of 'austerity and efficiency' are usually applied.[13,14] 'Efficiency' here mainly concerns the extent to which projects succeed in reducing risks of potential damage and casualties at the lowest possible cost.[15]

Investments in secondary objectives, such as improving spatial and environmental quality, are not eligible for financial support. Project proposals that focus on these aspects will therefore need to be funded in part through other means. A recent evaluation of the Water Act emphasises that excessive frugality 'can lead to additional costs as a result of resistance from the local community and delays in planning procedures'.[16] Insufficient consideration of social factors can thus hinder efficiency.

Within the Delta Programme for Spatial Adaptation, there is also scope to pay greater attention to social factors. Although the use of 'coupling opportunities' is explicitly part of the ambitions, it is notable that these opportunities are currently sought primarily in the energy transition, new construction and renovation.[17]

A different approach is possible. The Room for the River programme demonstrates that a dual societal objective—ensuring flood protection and improving spatial quality—can indeed be achieved (see also Chap. 3). At the municipal level, we are seeing smart combinations of physical and social

[9] Nationaal Deltaprogramma (2024b).

[10] Rijkswaterstaat (n.d.a).

[11] Ligtvoet and Knoop (2014).

[12] Ligtvoet and Knoop (2014).

[13] Rijkswaterstaat (n.d. b.).

[14] HWBP (2021).

[15] Board of Government Advisors (2020).

[16] Sweco Nederland et al. (2024: 18) .

[17] Climate Adaptation Netherlands (2020).

Box 6.1 (continued)

adaptation goals as well, for example in the Sprenkelaar district of Apeldoorn, where promoting social connection and interaction is one of the three priorities in area development. In other municipalities, the necessary replacement of sewer systems is being used as an opportunity to improve the social fabric of districts through spatial interventions. In a neighbourhood in The Hague, such a renovation includes increasing green space and creating areas for people to meet.[18]

Finally, the municipality of Dordrecht has investigated how higher-lying areas can serve as shelters during floods. In a new residential district, shelters are being provided for residents from other districts. The municipality is also raising awareness of flood risks, for example by organising a water festival.[19]

In a Climate-Resilient Society, People Look Out for Each Other, Trust Each Other and Have the Capacity to Act

We are at the beginning of a long period in which climate change will inevitably lead to climate events with profound physical, economic, ecological and societal consequences. While these effects may be beneficial for certain locations and groups, global data from—for example—the insurance sector clearly indicate that significant societal damage is and will continue to be incurred.[20] The Netherlands, too, must anticipate more frequent disruptive and potentially life-threatening extreme weather events. It is also possible that climate change will give rise to weather phenomena that currently fall entirely outside expectations for the Netherlands. At the same time, there will continue to be considerable uncertainty regarding where and when certain types of climate events will occur.

This combination of increased intensity and inherent uncertainty in climate developments and threats requires the anticipation of circumstances that the Dutch have rarely or never experienced before, and may not even be able to foresee (the 'unknown unknowns'). One example is cascading effects, for instance when a climate event leads to prolonged power outages, which in turn cause problems in drinking water supply, financial services and health care institutions. This alone means that the Netherlands will not be able to contain the impact of climate change in advance through technical or preventive measures. A climate-resilient society will need to look beyond such measures.

---

[18] Municipality of The Hague (2021: 14).

[19] Mecanoo Architecten (2023); Resilient Delta Initiative (2023).

[20] A recent study by an international reinsurer shows that the total amount of claims paid out in 2023 was higher than the average annual amount over the past ten years. Climate change is cited as one of the main causes; see Munich Re (2024).

The analyses in Chaps. 3, 4 and 5 show that this is most achievable in a society where people look out for each other, trust each other and have the capacity to act.[21] A recent report by the Netherlands Institute for Public Safety describes how resilience manifests itself during a crisis:

> A well-functioning and well-practiced social network shares knowledge, expertise and information, ensures monitoring and situational awareness, increases public preparedness and awareness through risk communication, maintains strategic reserves and capacities, and ensures coordinated deployment as well as effective recovery.[22]

Is the Netherlands climate-resilient from this perspective? Although social cohesion in the Netherlands is relatively stable and high—both horizontally (between people) and vertically (between citizens and institutions)—there are also significant differences between groups, and substantial segments of society to which this does not apply.[23]

In addition, official measures aimed at strengthening citizens' climate resilience currently do not reach everyone. Consider, for example, the subsidy schemes discussed in Chap. 5, which mainly benefit the relatively affluent. The availability of knowledge and skills within the social network also influences whether citizens take adaptive action in their daily environment and lifestyles.[24] In other words: if citizens are expected to take action themselves to increase their climate resilience, living in poverty and social isolation forms a significant barrier.

Especially in comparison to energy transition policy, it is notable that climate adaptation in the Netherlands—outside the domain of flood protection—is often characterised by voluntariness and individual responsibility. For example, existing homes with a low energy label may no longer be rented out from 2030, but there is limited regulation for tenants to rely on in the case of heat stress.[25] The authorities largely leave it to property owners to make homes climate-resilient. This policy approach increases the vulnerabilities of isolated or socio-economically disadvantaged households.

Data from Statistics Netherlands show that nearly half of these households are unable to cool their homes on hot days, compared to only a quarter of more socio-economically advantaged households.[26] Given that people with a low socio-economic position are more likely to live in rental housing, they are largely dependent on landlords to address this issue. This dependency results in a reduced capacity to act against the consequences of climate change, with all the negative effects that entails.[27]

---

[21] WRR (2017).

[22] NIPV (2024: 9).

[23] SCP (2024a).

[24] For various detailed examples, see Don (2025).

[25] For existing policy on the sustainability of the built environment and measures for heat stress, see BZK (2022); !wOON (n.d.).

[26] CBS (2024).

[27] For an extensive discussion of the consequences of a reduced capacity to act, see WRR (2023).

In Box 6.2, we discuss how a multitude of initiatives in Rotterdam is leading to greater societal action, with the authorities involved making smart use of knowledge about social factors. These examples may serve as inspiration for the rest of the Netherlands.

Box 6.2 Rotterdams WeerWoord: rallying everyone behind adaptation

For Rotterdam, climate adaptation is not a choice, but a necessity. 85 to 90% of the Dutch port city lies below sea level, making it highly vulnerable to the consequences of climate change. The Rotterdams WeerWoord 2030 programme has been developed to enhance the city's resilience. This programme implements a joint strategy aimed at ensuring that Rotterdam is fully climate-resilient by 2050.[28]

The first priority is the implementation of physical measures to make the city climate-resilient. Examples include the construction of water storage systems and infrastructure improvements that increase the city's resistance to heavy rainfall and flooding. Measures such as urban greening and improved water retention not only enhance climate resilience, but also contribute to residents' well-being.

A key element of the Rotterdam approach is the use of maps. The *Koele Kaart* (cold map), for instance, indicates the locations of cool public buildings and drinking water points. The map makes it easier for residents affected by heat stress to find relief. District climate maps provide detailed insights into vulnerabilities at the district level, such as vertical evacuation options and risks of water nuisance. This information enables the development of solutions tailored to the needs of each neighbourhood.

Rotterdam's residents are also encouraged to take action themselves. For example, the *Tegeltaxi* service helps residents who wish to green their surroundings by collecting removed paving stones. The toolkit—an online database containing dozens of physical and non-physical measures—offers specific perspectives for action. In the words of the programme:

> Would you like to help make Rotterdam climate-adaptive? Are you a professional city-maker, tenant, homeowner or entrepreneur? Climate adaptation concerns all Rotterdammers. We all have a part to play. It is helpful to have some guidance along the way. That is why WeerWoord Rotterdam has created this toolkit.

Social engagement plays a major role in WeerWoord. For example, a theatre group has developed a performance to raise climate awareness and offer practical solutions. The actors inform older people and their carers in an accessible and non-patronising way about how to cope with heat and what they can do to stay healthy during warm periods.

---

[28] Rotterdams WeerWoord (n.d.).

Box 6.2 (continued)

In addition, various neighbourhood initiatives have been launched to strengthen social cohesion by actively involving residents in climate adaptation projects. In this way, community spirit and climate preparedness go hand in hand.

6.4 Recommendations

We began this book with a description of an old Dutch winter landscape in a painting by Hendrick Avercamp. That painting reminds us that the climate and its associated weather conditions have a profound impact on the Dutch identity. Dykes and Delta works define the contours of our country. It would be naïve to think that increasing heat, drought, heavy rainfall and a greater risk of flooding would have no social consequences. We have shown that climate events inevitably leave social traces and that societal preparedness and resilience help determine how disruptive those traces will prove to be. We conclude that the pursuit of climate resilience can be significantly strengthened by paying greater attention to social factors than has so far been the case in Dutch policy.

With this in mind, the WRR makes three main recommendations for government policy:

- Invest in social infrastructure to promote societal resilience.
- Promote the joining of forces and mutual trust.
- Ensure that people know what they can do to adapt to a changing climate.

Recommendation 1: Invest in social infrastructure to promote societal resilience

- Combine physical safety and societal resilience ambitions when designing or redesigning inhabited areas.
- Prioritise the social infrastructure of communities that are physically and/or socio-economically most vulnerable to the effects of climate change.

In a free society, people themselves decide whether and how to interact with others, and with whom. However, authorities can make it easier for citizens to meet, connect, support each other and undertake joint initiatives. Investing in what has come to be known as social infrastructure—the physical places and conditions that facilitate encounters between people and enable them to develop social capital[29]—is the most

[29] Kolner and Engbersen (2024: 6); Klinenberg (2018).

important lever for this. Such infrastructure fulfils numerous functions that are of great importance to societal resilience in times of crisis. It provides an additional channel for reliable information provision, helps to identify problems at the neighbourhood level, has the potential to stimulate community initiatives and can play a role in recovery operations.[30] In Box 6.3, we take a closer look at social infrastructure.

Box 6.3 Social infrastructure

From metro stations to bus stops, from festivals to the weekly small-scale market on the village green, from large central libraries to the local brass band: social infrastructure is everywhere. Latham and Layton distinguish five different types of physical space that are part of it.[31] First, there are public places, such as museums, parks, libraries, squares and schools. These spaces are widely accessible and often wholly or partially funded by public resources. The second type of space comprises locations with a commercial purpose. Think of markets and cafés. The third category concerns recreational spaces—for example, public swimming pools or recreational lakes, sports clubs, stadiums and cultural institutions. Religious spaces, such as synagogues, churches and mosques, form the fourth category. The final category relates to transport infrastructure such as stations (and their surroundings) and pedestrian and bicycle paths. In addition, social infrastructure can also take non-physical forms, for example as online social platforms or civil-society organisations.[32]

The mere existence of these spaces does not mean that we use them in ways that contribute to social cohesion and societal resilience. Sometimes, the opposite is true: a poorly designed or maintained park, for example, can lead to feelings of insecurity, so the likelihood that people will use such a park to meet each other is low. If social infrastructures are not broadly accessible, there may be a risk of exclusion. For most functions that social infrastructure can serve, the involvement of people is indispensable. For example, providing information requires more than just distributing leaflets; a question-and-answer session or a help desk is often far more effective.

A recent report from Erasmus University Rotterdam shows how, during the coronavirus pandemic, staff at the public library in Rotterdam managed to use social infrastructure to enhance the city's resilience. They proactively reached out to older people registered with them to see if and how they could be helped. Sometimes they delivered groceries, while at other times they referred people to other organisations. Cooperation played an important role. Such partnerships can also make a difference in the field of climate adaptation, for example during periods of prolonged heat.

[30] These functions are derived from Kolner and Engbersen (2024).

[31] Latham and Layton (2019). We note that there is some overlap between the five different types.

[32] Kolner and Engbersen (2024).

Box 6.3 (continued)

In terms of spatial design, municipalities and provinces present a greener built environment as an opportunity to become both more climate resilient and to promote social infrastructure in neighbourhoods.[33] Scientific research supports the view that these local and regional authorities are pursuing a promising course. More greenery in a district reduces the impact of heatwaves and excessive rainfall. At the same time, it can promote social interactions, which in turn can benefit social cohesion and thus resilience in times of need. Spatial design initiatives of this kind deserve to be widely emulated.

However, this is no easy task. Recent research has shown that, in practice, greening in Dutch cities is declining and urban hardscaping is increasing. In the 32 largest municipalities, about 3.6 million people lived in a hardscapened environment in 2024.[34] There are still considerable obstacles to overcome in the implementation of projects that take both physical protection and social factors into account. This will require the skill and willingness to combine different policy areas and budgets.[35] It is also important that residents' wishes and knowledge about the existing social infrastructure are incorporated into designs. For example, if greenery displaces a much-used football pitch, this can have a counterproductive effect from a social perspective. Previous recommendations have pointed out that clear standards encouraging local authorities to achieve multiple benefits with urban nature development can be helpful. A requirement to make social infrastructure part of municipal environmental visions may also help.[36]

The government must set higher targets and maintain a faster pace when it comes to strengthening the social infrastructure in districts and regions where it is needed. Precisely because people in such districts often lack the capacity to take appropriate action in and around their own homes, there must be greater opportunities to seek protection in public spaces and buildings. For example, priority should be given in these districts to the installation of climate control systems in community centres, libraries, schools and health care institutions.

More generally, libraries, community centres, sports clubs, cultural societies and religious or spiritual groups are a vital part of the social infrastructure that connects people in the Netherlands. Their significance as meeting places, information channels and incubators for social trust extends far beyond their primary objective. From the perspective of climate adaptation, initiatives aimed at strengthening community cohesion therefore deserve greater support.

---

33 See City Deal for Public Space (2024); Province of Gelderland (2023); Municipality of Rotterdam (2023).

34 Sweco Netherlands (2024).

35 City Deal for Public Space (2024).

36 RVS (2022: 18); PBL (2023).

Recommendation 2: Promote the joining of forces and mutual trust

- Encourage parties in the social, spatial and security domains to commit to strengthening social infrastructure together.
- Use the methods and experiences of the Dutch Delta Programme to stimulate investment, knowledge sharing and cooperation in this area.

Given what is at stake and how much can be gained by joining forces, it is important to take a critical look at how policy is currently organised. Dutch policy documents and programme structures suggest that a cross-sectoral approach is being taken to tackling the broad challenge of adaptation. The Ministry of Infrastructure and Water Management coordinates the development of the National Climate Adaptation Strategy involving other ministries and local authorities. In practice, however, the current Dutch climate adaptation policy approach appears to be fragmented and budgets are compartmentalised. Moreover, this policy has traditionally been dominated by physical solutions.[37]

To incorporate social factors into adaptation measures effectively, greater cooperation will be required. In regional spatial development, for instance, budgets for physical and social measures could be combined. It is also important to consider that, from a national security perspective, efforts are being made to enhance societal resilience to threats from both state and non-state actors.[38] The same applies to pandemic preparedness and planning for other disasters. Meanwhile, parties within the social domain have long advocated for strengthening the 'social foundation' in districts and communities. In other words, there are various stakeholders and rationales for investing in social infrastructure.[39] However, there is a genuine risk of fragmentation of efforts or delays. This is another reason to prioritise the joining of forces. The Delta Programme can provide inspiration for how to begin combining these forces. The government-appointed commissioner ('Delta Programme Commissioner') acts as a nationally operating, connecting figure, with a long-term mandate, a direct line to the responsible minister and the House of Representatives and sufficient incentive funding to encourage the parties to share knowledge, work together and innovate. The systematic working towards and recalibration of guiding decisions has brought calm and regularity to policy development and led to the establishment of a large, diverse and sustainable 'Delta community'.[40]

Such a vibrant exchange of ideas and experiences is currently lacking when it comes to incorporating social factors into adaptation policy. As a result, successful local initiatives often remain hidden from the view of other municipalities or regions.

[37] The revision of the National Climate Adaptation Strategy also pays attention to social impact.

[38] WRR (2024). Climate adaptation measures can also be affected by these threats, especially in the case of digital vulnerability, see: WRR (2019); Netherlands Court of Audit 2019.

[39] Movisie and Verweij-Jonker (2024).

[40] Van Buuren and Teisman (2014); Van Popering-Verkerk et al., (2019).

Knowledge sharing and cooperation between government bodies and across domains is crucial to strengthening social infrastructure in all regions. This requires leadership that facilitates this knowledge sharing and cooperation and provides government bodies with incentives to invest in social infrastructure in a targeted manner. This will strengthen societal resilience. At the political level, too, this requires a joining of forces, for example in the form of a ministerial committee on climate resilience with representatives from the physical, social and administrative domains.

Recommendation 3: Ensure that people know what they can do to adapt to a changing climate

- Use risk and crisis communication that enhances everybody's capacity to act.
- Invest in the necessary data collection and monitoring.
- Make sure that the public facilities for people and businesses are widely known and accessible during a climate event.

Climate change is creating a new reality: threats from extreme weather are increasing, new and more unpredictable. They also cover a broader spectrum than the range of flood-related threats that the Dutch have historically experienced and learned to take seriously. This new reality demands that society deals with climate risks in a different way. Providing sound information about climate risks and perspectives for action is essential in this regard.

It is up to the authorities to provide information about risks that is as specific as possible. Increased data collection and monitoring will make it possible to enrich the data in the Climate Impact Atlas, which provides insight into risks and vulnerabilities at the neighbourhood level. There is still much to be gained in this area. The advisory group of the Delta Programme recently noted that the available information is not always suitable for an adequate assessment of the risks faced by households.[41] For example, building-specific characteristics are not taken into account, even though minor structural modifications can be highly decisive for exposure to threats. For instance, the presence of high doorsteps can prevent water damage during heavy rainfall. In addition, the quality of available data varies. Municipalities such as Amsterdam and Rotterdam map out specific foundation risks, whereas in other municipalities only national-level information is available.

However, increasing risk awareness alone is not enough. Only when people, businesses and communities feel a greater capacity to act when it comes to increasing their climate resilience will social change occur.[42] The Netherlands can learn from how other countries encourage their citizens to prepare for emergencies. Consider, for example, how in Japan citizens are urged to take earthquakes into account and

[41] Deltares (2024); Nationaal Deltaprogramma (2024a).

[42] Cf. WRR (2023).

how they are provided with specific perspectives for action.[43] Sometimes, practical guidelines may suffice, such as advice on the importance of removing standing water around the home to prevent mosquito populations.

It is also necessary for the authorities to provide insight into their risk assessments and the policy objectives they themselves pursue in this area. This requires the courage to be transparent. If the authorities are not transparent about which climate risks they are addressing, it is unrealistic to expect society to take action. While clear standards have been set in the area of flood protection, there are no specific objectives for dealing with other climate threats. As a consequence, climate resilience has only been defined for the flood protection domain.

In this book, we have shown that certain groups in society are particularly susceptible to the risks of climate change. People who are isolated or in a weak socio-economic position have less capacity to act—for example, to take measures to limit the consequences—while their exposure to certain types of climate events is greater. Consider, for example, heat issues in social housing. A societal approach to climate adaptation will prevent this exposure from becoming worse. This leads to several policy implications.

Policymakers should pay particular attention to the financial constraints that vulnerable households face in implementing climate adaptation measures. The self-reliance of citizens varies. More ambitious standards—for example, for housing associations—can be an effective way to make vulnerable groups better able to cope with climate threats. If the government decides to promote the purchase of emergency packages, it should be considered that such packages are not easily affordable for every citizen.

Increased financial vulnerability is not limited to households with a low socio-economic position. If market actors start to price in climate risks more explicitly in the housing market, this may lead to declines in property values for current homeowners or to unforeseen repair costs. For some households, foundation repair work will be difficult to afford. These households are also vulnerable. This calls for social policy choices in the housing sector.

How people cope with their losses and experiences during and after a climate event is partly dependent on how the authorities communicate and act during and after the acute phase of such an event. It is important to understand and respond to the circumstances and feelings of those affected. This offers citizens the best chance to regain control over their lives[44] and reduces the risk of social disruption during and after a climate event. For now, the Dutch ombudsman has observed that the authorities adopt a top-down approach in most recovery processes, which does not always align with what citizens need.[45]

[43] Whitney (2021). The community also often plays an important role in Japan in taking measures; see Nakai and Nakano (2023).

[44] SCP (2024b).

[45] Nationale Ombudsman (2023).

The Dutch government is already aware of the importance of clear communication and increasingly tailors its communication to vulnerable groups.[46] The sign language interpreter at the prime minister's press conferences during the coronavirus pandemic is an example of this. Nevertheless, the floods of the Geul river in 2021 revealed that improvements are possible, particularly in terms of clarity and inclusivity.[47] People with health problems, less financial means or a small social network have different needs during an evacuation. For risk communication as well, it is important that the message is sufficiently tailored to the experiences and perspectives of the intended target group, with particular attention to reaching and providing perspectives for action for the most vulnerable groups.

6.5 In Conclusion

Many Dutch people are proud of the centuries-old 'battle against the water' that has shaped the country and its inhabitants. This struggle and the associated pride are embedded in the cultural DNA and the shared identity of the Dutch.[48] In the future, will they be able to take equal pride in how they responded to climate threats as a society? The challenges are immense. We have demonstrated that a thorough understanding of the role of social factors in climate adaptation offers opportunities for strengthening climate policy. A climate adaptation policy that takes the interaction between physical and social dynamics as its starting point has much to offer. It provides the best route available to evolve towards a climate-resilient society. Especially now that the Kingdom of the Netherlands must learn to live with rising sea levels, extreme rainfall, drought, heat and hurricanes, a new opportunity for pride presents itself: pride in an approach to climate adaptation that not only offers better protection to residents, but also reconnects them.

References

!woon. (n.d.). *Te hoge binnentemperatuur—hittewoning*. https://www.wooninfo.nl/vraagbaak/onderhoud/hittewoning/

AD. (2022, November 29). Iedereen een noodpakket in huis, adviseert NCTV: 'Nederland extra kwetsbaar'. *Algemeen Dagblad*. https://www.ad.nl/politiek/iedereen-een-noodpak-ket-in-huis-adviseert-nctv-nederland-extra-kwetsbaar~ad468981/?cb=902ee1964954de3bb2ce5b0b5c25daaf&cb=2f8feb10ff22c0134260c44ab3b73ace&auth_rd=1&auth_rd=1

Algemene Rekenkamer. (2019). *Digitale dijkverzwaring: cybersecurity en vitale waterwerken*. https://www.rekenkamer.nl/publicaties/rapporten/2019/03/28/digitale-dijkverzwaring-cybersecurity-en-vitale-waterwerken

[46] NCTV (2024).

[47] Van den Berg et al., (2022: 34).

[48] Jensen (2024).

Algemene Rekenkamer. (2023). *Voorbij de dijk. Keuzes in het waterveiligheidsbeleid*. https://www.rekenkamer.nl/publicaties/rapporten/2023/10/12/voorbij-de-dijk

BZK. (2022). *Volkshuisvesting en Ruimtelijke Ordening: Beleidsprogramma Versnelling Verduurzaming Gebouwde Omgeving*. Ministerie van Binnenlandse Zaken en Koninkrijkrelaties. https://open.overheid.nl/documenten/ronl-789924103b28f6a32678bdd3fc81e5d35b2a320a/pdf

CBS. (2024). *1 op de 3 huishoudens krijgt woning niet koel op warme dagen*. Centraal Bureau voor de Statistiek. https://www.cbs.nl/nl-nl/nieuws/2024/29/1-op-de-3-huishoudens-krijgt-woning-niet-koel-op-warme-dagen

City Deal for Public Space / City Deal Openbare Ruimte. (2024). *Routekaart voor een integrale aanpak van stedelijke transitieopgaven door de City Deal Openbare Ruimte*. https://www.citydealopenbareruimte.nl/default.aspx

College van Rijksadviseurs. (2020). *Hoogwaterbeschermingsprogramma: van 'sober en doelmatig' naar 'slim en doelmatig'*. https://www.collegevanrijksadviseurs.nl/adviezen-publicaties/publicatie/2020/07/21/hoogwaterbeschermingsprogramma-van-sober-en-doelmatig-naar-slim-en-doelmatig

Deltares. (2024). *Signaal 3: Kansen en gevaren van klimaatrisicoinformatie op huisniveau*. https://www.deltaprogramma.nl/documenten/publicaties/2024/11/29/signaal-3.-kansen-en-gevaren-van-klimaatinformatie-op-huisniveau

Don, C. (2025). *Klimaatgetto's: hoe klimaatverandering onze huizen binnendringt*. Walburg Pers.

Dückers M. (2023). De rampen die we samen creëren: Terugkerende patronen in crisisaanpak. *De Psycholoog*, 60(5). https://www.tijdschriftdepsycholoog.nl/wetenschap/de-rampen-die-we-samen-creeren/

Municipality of The Hague / Gemeente Den Haag. (2021). *Visie herinrichting Bomenbuurt*. https://hoplrcontent.blob.core.windows.net/attachments/Visie%20herinrichting%20bomenbuurt_343116eebb0de0-0bb9-4938-85c6-f92e21646647.pdf

Municipality of Rotterdam / Gemeente Rotterdam. (2023). *Rotterdam gaat voor groen: de groenagenda 2023–2026* (23bb004220). https://gemeenteraad.rotterdam.nl/Agenda/Document/1d7e338f-56fd-4c1e-8246-deb5abbe8fe9?documentId=df6b2db4-9d28-468a-9ca4-56e579e3889d&agendaItemId=3a7c6852-5ea1-4e7c-9064-13ba53a0f8dd

HWBP. (2021). *Stappenplan beoordeling sober en doelmatig ontwerp*. Hoogwaterbeschermingsprogramma. https://www.hwbp.nl/kennisbank/sober-en-doelmatig/documenten/handreikingen/2021/10/25/stappenplan-sober-en-doelmatig

IenW. (2023). *Nationaal Uitvoeringsprogramma Klimaatadaptatie Slimmer, intensiever, voor en door iedereen*. https://www.rijksoverheid.nl/documenten/rapporten/2023/11/17/bijlage-2-nationaal-uitvoeringsprogramma-klimaatadaptatie-nup-ka

IenW. (2024). *Brief Water en Bodem naar aanleiding van het Tweeminutendebat Water 8 oktober 2024* [Kamerbrief]. Ministerie van Infrastructuur en Waterstaat. https://open.overheid.nl/documenten/dpc-d278a56998fb4253af9f35329e230b7bacafc217/pdf

Informatiepunt Leefomgeving. (n.d.). *Landelijke Coördinatiecommissies bij watercrises*. https://iplo.nl/thema/water/crisismanagement-water/landelijke-coordinatiecommissies/

Jensen, L. (2024). *Rampen: Een nieuwe geschiedenis van Nederland*. Uitgeverij Prometheus.

Klimaatadaptatie Nederland. (2020). *Handreiking Slim Koppelen klimaat-adaptatie voor gemeenten*. https://klimaatadaptatienederland.nl/hulpmiddelen/overzicht/handreiking-slim-koppelen/

Klinenberg, E. (2018). *Palaces for the people: How social infrastructure can help fight inequality, polarization, and the decline of civic life*. Crown.

Kolner, C., & Engbersen, G. (2024). Vitale sociale infrastructuur bereikt burgers en redt levens. *Sociale Vraagstukken*. https://www.socialevraagstukken.nl/vitale-sociale-infrastructuur-bereikt-burgers-en-redt-levens/

Latham, A., & Layton, J. (2019). Social infrastructure and the public life of cities: Studying urban sociality and public spaces. *Geography Compass, 13*(7), Article e12444.

Ligtvoet, W. Knoop, J. (2014). *Kleine kansen—grote gevolgen: Slachtoffers en maatschappelijke ontwrichting als focus voor waterveiligheidsbeleid*. https://www.pbl.nl/publicaties/kleine-kansen-grote-gevolgen

Mecanoo Architecten. (2023). *Maasterras Stedenbouwkundig Masterplan.*

Movisie & Verweij-Jonker. (2024). *De sociale basis versterken, investeren in de toekomst. Veranderthеorie voor overheden en maatschappelijke partners.* https://www.movisie.nl/publicatie/sociale-basis-versterken-investeren-toekomst

Munich Re. (2024). *Record thunderstorm losses and deadly earthquakes: the natural disasters of 2023.* https://www.munichre.com/en/company/media-relations/media-information-and-corporate-news/media-information/2024/natural-disaster-figures-2023.html#video

Nakai, F., & Nakano, G. (2023). Community-mediated individual disaster preparedness practices: A case study in Kochi, Japan. *International Journal of Disaster Risk Reduction, 86*, Article 103532.

Nationaal Deltaprogramma. (2022). *Advies Signaalgroep Deltaprogramma 2022.* https://www.deltaprogramma.nl/documenten/publicaties/2022/11/10/adviesbrief-signaalgroep-2022

Nationaal Deltaprogramma. (2024a). *Advies Signaalgroep Deltaprogramma 2024.* https://www.deltaprogramma.nl/documenten/publicaties/2024/11/29/advies-signaalgroep-deltaprogramma-2024

Nationaal Deltaprogramma. (2024b). *Nu voor later. Deltafonds* (hoofdstuk 07). https://minienw.sitearchief.nl/?subsite=dp2024deltaprogramma#archive

Nationale ombudsman. (2023). *Herstel bieden: een vak apart* (2023/157). https://www.nationaleombudsman.nl/system/files/bijlage/Herstel%20bieden_een%20vak%20apart.pdf

NCTV. (2024). *Handreiking Beter voorbereid op rampen en crises.* Nationaal Coördinator Terrorismebestrijding en Veiligheid. https://www.nctv.nl/documenten/publicaties/2024/05/23/handreiking-beter-voorbereid-op-rampen-en-crises

Netherlands Court of Audit / Nationale ombudsman. (2023). *Herstel bieden: een vak apart* (2023/157). https://www.nationaleombudsman.nl/system/files/bijlage/Herstel%20bieden_een%20vak%20apart.pdf

NIPV. (2024). *Versterken van veerkracht. Naar een gezamenlijke aanpak van ongekende crises.* Nationaal Instituut Publieke Veiligheid. https://nipv.nl/wp-content/uploads/2024/05/20240501-NIPV-Crisisplan-Versterken-van-veerkracht.pdf

NOS. (2022, November 29). *Nederlanders niet goed voorbereid op ramp: 'We zijn een kwetsbaar land'.* Nederlandse Omroep Stichting. https://nos.nl/artikel/2454384-nederlanders-niet-goed-voorbereid-op-ramp-we-zijn-een-kwetsbaar-land

NRC. (2024a, January 12). *De minister gaat het verbieden, maar Arnhem bout nog wél in de uiterwaarden.* https://www.nrc.nl/nieuws/2024/01/12/de-minister-gaat-het-verbieden-maar-arnhem-bouwt-nog-wel-in-de-uiterwaar-den-a4186831

NRC. (2024b, July 8). *Achtduizend woningen in een van de diepste polders. Kan dat wel veilig?* https://www.nrc.nl/nieuws/2024/07/08/kan-het-dorp-in-die-diepe-polder-wel-veilig-robuust-en-klimaatbestendig-worden-gebouwd-a4859079

PBL. (2023). *Natuur in en om de stad* (5023). Planbureau voor de Leefomgeving. https://www.pbl.nl/uploads/default/downloads/pbl-2023-natuur-in-en-om-de-stad-5023.pdf

PBL. (2024). *Monitor nationale omgevingsvisie 2024: tweede vervolgmeting* (5313). Planbureau voor de Leefomgeving. https://www.pbl.nl/system/files/document/2024-09/pbl-2024-monitor-nationale-omgevingsvisie-5313.pdf

Provincie Gelderland. (2023). *Uitvoeringsagenda klimaatadaptatie 2024–2027*. https://media.gelderland.nl/Uitvoeringsagenda_Klimaatadaptatie_2024_2027_dc8ac817a7.pdf

Resilient Delta Initiative. (2023). *Integrated Action Framework.* Towards a safe and climate-adaptive Maasterras.

Rijkswaterstaat. (n.d.a). *Ruimte voor de rivieren.* https://www.rijkswaterstaat.nl/water/waterbeheer/bescherming-tegen-het-water/maatregelen-om-overstromingen-te-voorkomen/ruimte-voor-de-rivieren

Rijkswaterstaat. (n.d. b). *Waterprojecten.* https://www.rwseconomie.nl/werkwijzers/mkba-bij-mirt-verkenningen/bepalen-effecten-baten/waterprojecten

Rli. (2024). *Ruimtelijke ordening in een veranderend klimaat*. Raad voor de leefomgeving en infrastructuur. https://www.rli.nl/sites/default/files/Rli%202024-02%20Klimaatadaptatie%20eindversie.pdf

Rotterdams WeerWoord. (n.d.). *Rotterdams WeerWoord*. Gemeente Rotterdam. https://rotterdamsweerwoord.nl

RVS. (2022). *Ruimte maken voor ontmoeting*. Raad voor Volksgezondheid en Samenleving. https://www.raadrvs.nl/adviezen/ruimte-voor-ontmoeting/documenten/publicaties/2022/09/06/ruimte-maken-voor-ontmoeting

SCP. (2024a). *Factsheet Hoe staat het ervoor met de sociale cohesie in Nederland?* Sociaal en Cultureel Planbureau. https://www.scp.nl/publicaties/publicaties/2024/06/04/hoe-staat-ervoor-met-de-sociale-cohesie-in-nederland

SCP. (2024b). *Kennisnotitie Overheidshandelen tijdens en na crises: Vier aandachtspunten voor goed overheidshandelen rond crisissituaties*. Sociaal en Cultureel Planbureau. https://www.scp.nl/publicaties/publicaties/2024/07/18/kennisnotitie-overheidshandelen-tijdens-en-na-crises

SCP. (2025). *Klimaat en Samenleving—Burgerperspectieven*. Sociaal en Cultureel Planbureau. https://www.scp.nl/publicaties/publicaties/2025/04/17/klimaat-en-samenleving

Sweco Nederland, TwynstraGudde, & Decisio. (2024). *Evaluatie wijzigingen ten aanzien van waterveiligheid in de Waterwet in 2014 en 2017, 2024*. https://open.overheid.nl/documenten/dpc-4f9ae16da1f6bfccd46faa746427fb14a11f4de8/pdf

Sweco Nederland. (2024). *Tekort aan groen in Nederlandse steden*. https://www.sweco.nl/wp-content/uploads/sites/5/2024/12/Versteningsonderzoek-NM-Sweco-2024.pdf

Terpstra, T. (2010). *Flood preparedness: Thoughts, feelings and intentions of the Dutch public*. University of Twente. https://doi.org/10.3990/1.9789036529549

Terpstra, T. (2011). Emotions, trust, and perceived risk: Affective and cognitive routes to flood preparedness behavior. *Risk Analysis, 31*(10), 1658–1675. https://doi.org/10.1111/j.1539-6924.2011.01616.x

Van Buuren, A., & Teisman, G. (2014). *Samen verder werken aan de Delta: de governance van het Deltaprogramma na 2014*. Erasmus Universiteit Rotterdam.

Van den Berg, K., Kolen, B., & Rudolph, M. (2022). *Evacuatiegedrag van getrof- fenen tijdens de overstromingen in Limburg in juli 2021* (PR3591.60). HKV Lijn in water. https://open.rijkswaterstaat.nl/open-overheid/onderzoeksrapporten/@257015/evacuatiegedrag-getroffenen-tijdens/

Van Popering-Verkerk, J., Teisman, G., & Van Buuren, A. (2019). *Eb en vloed van het Deltaprogramma: De Deltacommunity anno 2019 in beeld gebracht*. Erasmus University Rotterdam.

Van Thiel, L., & Mol, P. (2020). *Waterbewustzijn van de Nederlandse bevolking*. In opdracht van Kantar en HWBP. https://klimaatadaptatienederland.nl/@237838/waterbewustzijn-van-de-nederlandse-bevolking/

Whitney, J. (2021). *A new mantra for Japanese disaster management*. https://eastasiaforum.org/wp-content/uploads/2023/12/V.13-N.3.pdf

WRR. (2017). *Weten is nog geen doen. Een realistisch perspectief op redzaamheid* (rapport nr.97). Wetenschappelijke Raad voor het Regeringsbeleid. https://www.wrr.nl/publicaties/rapporten/2017/04/24/weten-is-nog-geen-doen

WRR. (2019). *Voorbereiden op digitale ontwrichting* (rapport nr.101). Wetenschappelijke Raad voor het Regeringsbeleid. https://www.wrr.nl/adviesprojecten/digitale-ontwrichting

WRR. (2023). *Grip: Het maatschappelijk belang van persoonlijke controle* (rapport nr.108). Wetenschappelijke Raad voor het Regeringsbeleid. https://www.wrr.nl/publicaties/rapporten/2023/11/30/grip

WRR. (2024). *Nederland in een fragmenterende wereldorde* (rapport nr.109). Wetenschappelijke Raad voor het Regeringsbeleid. https://www.wrr.nl/adviesprojecten/kantelende-wereldorde/documenten/rapporten/2024/07/01/nederland-in-een-fragmenterende-wereldorde

Glossary

Abbreviation	Term	Description
AFM	Dutch Authority for the Financial Markets (*Autoriteit Financiële Markten*)	Financial markets regulator
AIV	Advisory Council on International Affairs (*Adviesraad Internationale Vraagstukken*)	Government advisory body on foreign and international affairs
ANV	The National Network of Safety and Security Analysts (*Analisten Netwerk Veiligheid*)	Network of experts on safety and security issues
BZK	Ministry of the Interior and Kingdom Relations (*Ministerie van Binnenlandse Zaken en Koninkrijkrelaties*)	Ministry responsible for governance, democracy and Kingdom relations
CBS	Statistics Netherlands (*Centraal Bureau voor de Statistiek*)	Statistical office of the Netherlands
CPB	Netherlands Bureau for Economic Policy Analysis (*Centraal Planbureau*)	Independent government institute for economic policy
EU	European Union (*Europese Unie*)	Supranational political and economic union of European member states
EEA	European Environment Agency (*Europees Milieuagentschap*)	European agency providing environmental data and assessments
HWBP	Flood Protection Programme (*Hoogwaterbeschermingsprogramma*)	National programme for strengthening flood defences
IenW	Ministry of Infrastructure and Water Management (*Ministerie van Infrastructuur en Waterstaat*)	Ministry responsible for infrastructure, mobility and water management
Inspectie JenV	Inspectorate of Justice and Security (*Inspectie Justitie en Veiligheid*)	Supervisory body overseeing justice and security organisations
IPCC	Intergovernmental Panel on Climate Change	United Nations body assessing climate science

(continued)

R. El-Dardiry et al., *People and Climate*, Research for Policy,
https://doi.org/10.1007/978-3-032-25831-1

(continued)

Abbreviation	Term	Description
JenV	Ministry of Justice and Safety (*Ministerie van Justitie en Veiligheid*)	Ministry responsible for justice, policing and security
KCAF	Knowledge Centre for Foundation Problems (*Kennis Centrum Aanpak Funderingsproblematiek*)	Expertise centre on building foundation damage
KNMI	Royal Netherlands Meteorological Institute (*Koninklijke Nederlands Meteologisch Instituut*)	Dutch national meteorological institute
LNV	Ministry of Agriculture, Nature and Food Quality (*Ministerie van Landbouw, Natuur en Voedselkwaliteit*)	Former ministry responsible for agriculture, nature and food systems (now LVVN)
MNP	Netherlands Environmental Assessment Agency (*Milieu- en Natuuurplanbureau*)	Former environmental policy assessment institute (now PBL)
NAS	(Dutch) National Climate Adaptation Strategy (*Nationale klimaatadaptatiestrategie*)	National framework for climate adaptation policy
NCTV	National Coordinator for Counterterrorism and Security (*National Coördinator Terrorismebestrijding en Veiligheid*)	Government body coordinating counterterrorism and national security policy
NIPV	Netherlands Institute for Public Safety (*Nederlands Instituut Publieke Veiligheid*)	Knowledge institute for crisis management and emergency services
NJI	Netherlands Youth Institute (*Nederlands Jeugdinstituut*)	Dutch knowledge centre for youth and family policy
NOS	Dutch Broadcasting Foundation (*Nederlandse Omroep Stichting*)	Dutch public news broadcaster
NVM	Netherlands Association of Real Estate Agents and Appraisers (*Nederlandse vereniging van Makelaars*)	Association for real estate agents and appraisers in the Netherlands
OCW	Ministry of Education, Culture and Science (*Ministerie van Onderwijs, Cultuur en Wetenschap*)	Ministry responsible for education, culture and science
PBL	PBL Netherlands Environmental Assessment Agency (*Planbureau voor de leefomgeving*)	Independent government institute for environmental and spatial analysis
RIVM	National Institute for Public Health and the Environment (*Rijksinstituut voor Volksgezondheid en Milieu*)	Independent government knowledge organisation on public health and environmental risks
Rli	Council for the Environment and Infrastructure (*Raad voor de leefomgeving en infrastructuur*)	Independent government advisory council on environment and infrastructure to the Dutch government and parliament

(continued)

(continued)

Abbreviation	Term	Description
RVO	Netherlands Enterprise Agency (*Rijksdienst voor Ondernemend Nederland*)	Executive body of the ministry of Economic Affairs
RVS	Council for Public Health and Society (*Raad voor Volksgezondheid en Samenleving*)	Independent advisory body on public health and social issues
SCP	Netherlands Institute for Social Research (*Sociaal en Cultureel Planbureau*)	Research institute on social and cultural aspects of government policy
SER	Social and Economic Council (*Sociaal-Economische Raad*)	Advisory body on social and economic policy to the Dutch government and parliament
SES scores	Socio-economic status scores	Composite indicator of social and economic position
TNO	Netherlands Organisation for Applied Scientific Research (TNO)	Applied research organisation
VenW	Ministry of Transport, Public Works and Water Management (*Ministerie van Verkeer en Waterstaat*)	Former ministry responsible for transport, public works and water management (now IenW)
WKR	Scientific Climate Council (*Wetenschappelijke Klimaatraad*)	Advisory research body on climate policy to the Dutch government and parliament
WRR	Netherlands Scientific Council for Government Policy (*Wetenschappelijke Raad voor het Regeringsbeleid*)	Independent government scientific advisory body for the Dutch government

Bibliography

!woon. (n.d.). *Te hoge binnentemperatuur—Hittewoning*. https://www.wooninfo.nl/vraagbaak/onderhoud/hittewoning/

3Doc. (2022). *De klimaatverkenner.* Nederlandse Publieke Omroep (NPO). https://npo.nl/npo3/3doc-de-klimaatverkenner/28-10-2022/VPWON_1340943

ABN AMRO. (2023). *Stapeling klimaatrisico's en financiële draagkracht op de woningmarkt*. https://www.abnamro.com/research/nl/onze-research/stapeling-klimaatrisicos-en-financiele-draagkracht-op-de-woningmarkt

Abu Samah, A., & Shaffril, H. A. M. (2020). A comparative study between mainland and islander small-scale fishermen's adaptation towards climate change. *Environmental Science and Pollution Research, 27*, 11277–11289. https://doi.org/10.1007/s11356-019-07143-1

AD. (2022, November 29). Iedereen een noodpakket in huis, adviseert NCTV: 'Nederland extra kwetsbaar'. *Algemeen Dagblad*. https://www.ad.nl/politiek/iedereen-een-noodpak-ket-in-huis-adviseert-nctv-nederland-extra-kwetsbaar~ad4689981/?cb=902ee1964954de3bb2ce5b0b5c25daaf&cb=2f8feb10ff22c0134260c44ab3b73ace&auth_rd=1&auth_rd=1

Adélaïde, L., Chanel, O., & Pascal, M. (2022). Health effects from heat waves in France: An economic evaluation. *The European Journal of Health Economics, 23*, 1–13.

Agrawal, S., & Monroe, M. C. (2006). Using and improving social capital to increase community preparedness for wildfire. In USDA, *The Public and Wildland Fire Management: Social Science Findings for Managers* (pp. 163–167). US Department of Agriculture, Forest Service.

AIV. (2020). *Veiligheid en rechtsorde in het Caribisch gebied*. Den Haag: Adviesraad Internationale Vraagstukken. https://pure.rug.nl/ws/portalfiles/portal/210770200/Veiligheid_en_rechtsorde_in_het_Caribisch_gebied_AIV_advies_116_202009.pdf

Aldrich, D. P. (2012a). *Building resilience: Social capital in post-disaster recovery*. University of Chicago Press.

Aldrich, D. P. (2012b). Social, not physical, infrastructure: the critical role of civil society after the 1923 Tokyo earthquake. *Disasters, 36*(3), 398–419. https://doi.org/10.1111/j.1467-7717.2011.01263.x

Aldrich, D. P., & Meyer, M. A. (2015). Social capital and community resilience. *American Behavioral Scientist, 59*(2), 254–269.

Algemene Rekenkamer. (2019). Digitale dijkverzwaring: cybersecurity en vitale waterwerken. https://www.rekenkamer.nl/publicaties/rapporten/2019/03/28/digitale-dijkverzwaring-cybersecurity-en-vitale-waterwerken

Algemene Rekenkamer. (2023). *Voorbij de dijk. Keuzes in het waterveiligheidsbeleid*. https://www.rekenkamer.nl/publicaties/rapporten/2023/10/12/voorbij-de-dijk

Anderson, C. A. (2001). Heat and violence. *Current Directions in Psychological Science, 10*(1), 33–38. https://doi.org/10.1111/1467-8721.00109

R. El-Dardiry et al., *People and Climate*, Research for Policy,
https://doi.org/10.1007/978-3-032-25831-1

Anderson, C. A., Bushman, B. J., & Groom, R. W. (1997). Hot years and serious and deadly assault: Empirical tests of the heat hypothesis. *Journal of Personality and Social Psychology, 73*(6), 1213–1223. https://doi.org/10.1037//0022-3514.73.6.1213

Arcaya, M., Raker, E. J., & Waters, M. C. (2020). The social consequences of disasters: Individual and community change. *Annual Review of Sociology, 46*(1), 671–691.

Ariaans, J. (2024a, oktober 22). CDA Gennep wil opheldering: kan er nog gebouwd worden in Lob van Gennep? *AD.* www.ad.nl/gennep/cda-gennep-wil-opheldering-kan-er-nog-ge-bouwd-worden-in-lob-van-gennep~a1bfbb22/

Ariaans, J. (2024b, February 7). Inwoner Lob van Gennep moet huis zelf 'waterbestendig' maken. *De Gelderlander.*

Arnall, A., & Hilson, C. (2023). Climate change imaginaries: Representing and contesting sea level rise in Fairbourne, North Wales. *Political Geography, 102*, 102839, 1–9. https://doi.org/10.1016/j.polgeo.2023.102839

Atlas Leefomgeving. (2020). *Stedelijk hitte-eiland effect (UHI).* https://www.atlasleefomgeving.nl/stedelijk-hitte-eiland-effect-uhi

Australian Catholic Bishops Conference (ACBC). (2019). *The Catholic bishops of Australia have announced a National Prayer Campaign for the gift of rain during November.* https://www.dow.org.au/wp-content/uploads/2019/10/Bulletin-Notices-OOB-26-27-October-2019-WEB.pdf

Avercamp, H. (ca. 1620). *IJsvermaak bij een stad (Winterlandschap bij een stad)* [Olieverf op paneel]. Rijksmuseum. https://id.rijksmuseum.nl/200529588

Baart, T., & Schaap, T. (2007). *Bouwplaats Enschede: Een stad herschept zichzelf*. NAi Uitgevers.

Ballester, J., Quijal-Zamorano, M., Méndez Turrubiates, R. F., Pegenaute, F., Herrmann, F. R., Robine, J. M., Basagaña, X., Tonne, C., Antó, J. M., & Achebak, H. (2023). Heat-related mortality in Europe during the summer of 2022. *Nature Medicine, 29*(7), 1857–1866. https://doi.org/10.1038/s41591-023-02419-z

Bateman, T. S., & O'Connor, K. (2016). Felt responsibility and climate engagement: Distinguishing adaptation from mitigation. *Global Environmental Change, 41*, 206–215. https://doi.org/10.1016/j.gloenvcha.2016.11.001

Bazen, J. C. (2014). *We designed this neighborhood! Neighborhood satisfaction of local non-gentrifying residents with the reconstructed Roombeek neighborhood in Enschede, The Netherlands.* Utrecht University. https://studenttheses.uu.nl/handle/20.500.12932/17837

Bechtoldt, M. N., Götmann, A., Moslener, U., & Pauw, W. P. (2020). Addressing the climate change adaptation puzzle: A psychological science perspec- tive. *Climate Policy, 21*(2), 186–202. https://doi.org/10.1080/14693062.2020.1807897

Beck, J. V., Cole, K. D., Haji-Sheikh, A., & Litkouhl, B. (1992). *Heat conduction using Green's function.* Taylor & Francis.

Bihari, M., & Ryan, R. (2012). Influence of social capital on community preparedness for wildfires. *Landscape and Urban Planning, 106*(3), 253–261. https://doi.org/10.1016/j.landurbplan.2012.03.011

Birkland, T. A. (2006). *Lessons of disaster: Policy change after catastrophic events.* Georgetown University Press.

BNNVARA. (2017). ZEMBLA onthult: Dertig procent van bevolking Sint Maarten kampt met posttraumatische stressklachten na Irma. https://www.bnnvara.nl/zembla/artikelen/dertig-procent-van-bevolking-sint-maarten-kampt-met-posttraumatische-stressklachten-na-irma

Boin, A., Ekengren, M., & Rhinard, M. (2021). *Understanding the creeping crisis.* Springer Nature.

Bolin, B., & Kurtz, L. C. (2018). Race, Class, Ethnicity, and Disaster Vulnerability. In H. Rodríguez, W. Donner, & J. E. Trainor (Eds.), *Handbook of Disaster Research* (pp. 181–203). Springer International Publishing. https://doi.org/10.1007/978-3-319-63254-4_10

Bonfanti, R. C., Oberti, B., Ravazzoli, E., Rinaldi, A., Ruggieri, S., & Schimmenti, A. (2024). The Role of Trust in Disaster Risk Reduction: A Critical Review. *International Journal of Environmental Research and Public Health, 21*(1). https://doi.org/10.3390/ijerph21010029

Bos, C. K., Ullberg, S., & 't Hart, P. (2005). The Long Shadow of Disaster: Memory and Politics in Holland and Sweden. *International Journal of Mass Emergencies & Disasters*, *23*(1), 5–26. https://doi.org/10.1177/028072700502300101

Both, J. (n.d.). *De Wederopbouw op Goeree-Overflakkee*. https://geschiedenisvanzuidholland.nl/verhalen/verhalen/de-wederopbouw-op-goeree-overflakkee/

Botzen, W. J. W. (2017). *Economie van klimaatverandering en natuurrampen*. VU.

Bouman, T., Verschoor, M., Albers, C. J., Böhm, G., Fisher, S. D., Poortinga, W., Whitmarsh, L., & Steg, L. (2020). When worry about climate change leads to climate action: How values, worry and personal responsibility relate to various climate actions. *Global Environmental Change*, *62*, 102061. https://doi.org/10.1016/j.gloenvcha.2020.102061

Brock, A., Kwakernaak, M., de Meere, F., & Boutellier, H. (2019). *Literatuurstudie Sociale Cohesie. Ten behoeve van de ontwikkeling van de Maatschappelijke Impact Analyseren (MIA)*. Verwey Jonkers Instituut.

Brügger, A., Morton, T. A., & Dessai, S. (2015). Hand in hand: Public endorsement of climate change mitigation and adaptation. *PLoS ONE, 10*(4), Article e0124843. https://doi.org/10.1371/journal.pone.0124843

Buijs, J., Terpstra, T., Heuvel, J., Roest, A., Boogaard, F., Van der Maas, P., Fliervoet, J., Witteveen, L., Baan, F., & Heikoop, R. (2021). *Burgerparticipatie in klimaatadaptatie*. https://steenbreek.nl/wp-content/uploads/2022/01/Eindrapport-Burgerparticipatie-in-Klimaatadaptatie.pdf

BZK. (2022). *Volkshuisvesting en Ruimtelijke Ordening: Beleidsprogramma Versnelling Verduurzaming Gebouwde Omgeving*. Ministerie van Binnenlandse Zaken en Koninkrijkrelaties. https://open.overheid.nl/documenten/ronl-789924103b28f6a32678bdd3fc81e5d35b2a320a/pdf

BZK. (2023a). *Aanbieding Landelijke maatlat voor een groene klimaatadaptieve gebouwde omgeving en verkenning groennorm*. Ministerie van Binnenlandse Zaken en Koninkrijkrelaties. https://open.overheid.nl/documenten/ronl-cba8e18abd991aba311834eb957f270aa3081c80/pdf

BZK. (2023b). *Hittestress: Er is meer aandacht, maar een eenduidige aanpak ontbreekt nog*. Ministerie van Binnenlandse Zaken en Koninkrijksrelaties. https://www.klimaatweb.nl/wp-content/uploads/po-assets/876118.pdf

Cai, W., Wang, G., Dewitte, B., Wu, L., Santoso, A., Takahashi, K., Yang, Y., Carréric, A., & McPhaden, M. J. (2018). Increased variability of eastern Pacific El Niño under greenhouse warming. *Nature, 564*(7735), 201–206.

Cannavò, P. F. (2008). In the Wake of Katrina: Climate Change and the Coming Crisis of Displacement. *Political Theory and Global Climate Change*, 177–200. http://ndl.ethernet.edu.et/bit-stream/123456789/29172/1/8.pdf.pdf#page=204

Cannon, C., Fox Gotham, K., Lauve-Moon, K., & Powers, B. (2020). The climate change double whammy: Flood damage and the determinants of flood insurance coverage, the case of post-Katrina New Orleans. *Climate Risk Management, 27*, Article 100210. https://doi.org/10.1016/j.crm.2019.100210

Cannon, T. (2015). Disasters, climate change and the significance of 'culture.' In H. Krüger, G. Bankoff, T. Cannon, B. Orlowski, & E. L. F. Schipper (Eds.), *Cultures and Disasters: Understanding Cultural Framings in Disaster Risk Reduction* (pp. 88–106). Routledge.

Canon of the Netherlands / Canon van Nederland. (n.d.). *De stinkende stad ontvluchten*. https://www.canonvannederland.nl/nl/noord-holland/onderwijscanon-amstelland-en-de-meerlanden/de-stin-107984706

Caribisch Netwerk. (2021). *Klimaatverandering voor Bonairiaan ver van bed show: 'armoede overheerst'*. https://caribischnetwerk.ntr.nl/2021/10/25/klimaatverandering-voor-bonairiaan-ver-van-bed-show-armoe-de-overheerst/

Carleton, T. A., & Hsiang, S. M. (2016). Social and economic impacts of climate. *Science*, *353*(6304), aad9837.

Carlin, R. E., Love, G. J., & Zechmeister, E. J. (2014). Trust Shaken: Earthquake Damage, State Capacity, and Interpersonal Trust in Comparative Perspective. *Comparative Politics*, *46*(4), 419–437. http://www.jstor.org/stable/43664117

Castricum, R. (presentator). (2020). De Hofbar. De Lop van Gennep afvoerputje van Nederland? https://www.youtube.com/watch?v=VE6zYTZZtZA

CBS. (2018). *Sociale cohesie (binding met de buurt)*. Centraal Bureau voor de Statistiek. https://www.cbs.nl/nl-nl/nieuws/2018/01/tevredener-met-het-leven-bij-sterkere-binding-met-buurt/sociale-cohesie—binding-met-de-buurt

CBS. (2019). *Hogere sterfte tijdens recente hittegolf*. Centraal Bureau voor de Statistiek. https://www.cbs.nl/nl-nl/nieuws/2019/32/hogere-sterfte-tijdens-recente-hittegolf

CBS. (2020). *Sociale uitsluiting in Nederland: wie staat aan de kant*? Centraal Bureau voor de Statistiek. https://www.cbs.nl/nl-nl/longread/statistische-trends/2020/sociale-uitsluiting-in-nederland-wie-staat-aan-de-kant-/1-inleiding

CBS. (2021a). *Berekenwijze Sociaal Economische Status scores*. Centraal Bureau voor de Statistiek. https://www.cbs.nl/nl-nl/maatwerk/2021/45/berekenwijze-ses-score-per-wijk-buurt

CBS. (2021b). *Klimaatverandering en energietransitie: opvattingen en gedrag van Nederlanders in 2020*. Centraal Bureau voor de Statistiek. https://www.cbs.nl/nl-nl/longread/rapportages/2021/klimaatverandering-en-energietransitie-opvattingen-en-gedrag-van-nederlanders-in-2020?onepage=true#c-2—Opvattingen-over-klimaatverandering

CBS. (2022a). *Statusscore per wijk en buurt o.b.v. welvaart, opleidingsniveau en arbeid*. Centraal Bureau voor de Statistiek. https://www.cbs.nl/nl-nl/achtergrond/2022/14/statusscore-per-wijk-en-buurt-o-b-v-welvaart-opleidingsniveau-en-arbeid

CBS. (2022b). *Vertrouwen in Tweede Kamer steeg in 2020 en daalde in 2021*. Centraal Bureau voor de Statistiek. https://www.cbs.nl/nl-nl/nieuws/2022/13/vertrouwen-in-tweede-kamer-steeg-in-2020-en-daalde-in-2021

CBS. (2023a). *Klimaatverandering en energietransitie: Opvattingen en gedrag van Nederlanders in 2023*. Centraal Bureau voor de Statistiek. https://longreads.cbs.nl/klimaatverandering-en-energietransitie-2023/

CBS. (2023b). *Bijna 70 procent van ouderen heeft minstens elke week contact met de buren*. Centraal Bureau voor de Statistiek. https://www.cbs.nl/nl-nl/nieuws/2023/38/bijna-70-procent-van-ouderen-heeft-minstens-elke-week-contact-met-de-buren

CBS. (2023c). *Oversterfte in juni*. Centraal Bureau voor de Statistiek. https://www.cbs.nl/nl-nl/nieuws/2023/28/oversterfte-in-juni

CBS. (2024a). *Vrijwilligerswerk 2023*. Centraal Bureau voor de Statistiek. https://www.cbs.nl/nl-nl/longread/rapportages/2024/vrijwilligerswerk-2023?onepage=true#c-1—Inleiding

CBS. (2024b). *Sociale contacten en maatschappelijke participatie*. Centraal Bureau voor de Statistiek. https://opendata.cbs.nl/#/CBS/nl/dataset/85541NED/table?dl=922F0

CBS. (2024c). *Eenzaamheid; persoonskenmerken*. Centraal Bureau voor de Statistiek. https://opendata.cbs.nl/statline/#/CBS/nl/dataset/85766NED/table?ts=1731315091334

CBS. (2024d). *1 op de 3 huishoudens krijgt woning niet koel op warme dagen*. Centraal Bureau voor de Statistiek. https://www.cbs.nl/nl-nl/nieuws/2024/29/1-op-de-3-huishoudens-krijgt-woning-niet-koel-op-warme-dagen

CBS. (2024e). *Veiligheidsmonitor 2023*. Centraal Bureau voor de Statistiek. https://www.cbs.nl/nl-nl/longread/rapportages/2024/veiligheidsmonitor-2023/1-inleiding-en-samenvatting

CBS. (2024f). *Vertrouwen in mensen, recht en politiek; Europese vergelijking*. Centraal Bureau voor de Statistiek. https://www.cbs.nl/nl-nl/cijfers/detail/80518ned?dl=1EB57

CBS. (2025). *Leefstijl en sociaaleconomische status*. Centraal Bureau voor de Statistiek. https://www.cbs.nl/nl-nl/longread/statistische-trends/2025/leefstijl-en-sociaaleconomische-status

CBS. (n.d.). *Hoe interpreteer je de SES-WOA-scores en hoe zijn deze bepaald*? Centraal Bureau voor de Statistiek. https://www.cbs.nl/nl-nl/faq/infoservice/hoe-interpreteer-je-de-ses-woa-scores-en-hoe-zijn-deze-bepaald

Chakraborty, T., Hsu, A., Manya, D., & Sheriff, G. (2019). Disproportionately higher exposure to urban heat in lower-income neighborhoods: a multi-city perspective. *Environmental Research Letters*, 14, 105003. https://doi.org/10.1088/1748-9326/ab3b99/meta

Chamlee-Wright, E., & Storr, V. H. (2009). Club goods and post-disaster com-munity return. *Rationality and Society*, *21*(4), 429–458. https://ssrn.com/abstract=1711318

Chan, J., To, H.-P., & Chan, E. (2006). Reconsidering social cohesion: Developing a definition and analytical framework for empirical research. *Social Indicators Research, 75*(2), 273–302. https://doi.org/10.1007/s11205-005-2118-1

City Deal for Public Space / City Deal Openbare Ruimte. (2024). *Routekaart voor een integrale aanpak van stedelijke transitieopgaven door de City Deal Openbare Ruimte.* https://www.citydealopenbareruimte.nl/default.aspx

Cole-Dai, J. (2010). Volcanoes and climate. *WIREs Climate Change, 1*(6), 824–839. https://doi.org/10.1002/wcc.76

College van Rijksadviseurs. (2020). *Hoogwaterbeschermingsprogramma: van 'sober en doelmatig' naar 'slim en doelmatig'.* https://www.collegevanrijksadviseurs.nl/adviezen-publicaties/publicatie/2020/07/21/hoogwaterbeschermingsprogramma-van-sober-en-doelmatig-naar-slim-en-doelmatig

Cologna, V., & Siegrist, M. (2020). The role of trust for climate change mitigation and adaptation behaviour: A meta-analysis. *Journal of Environmental Psychology, 69*, Article 101428.

Comerio, M. C., Landis, J. D., & Rofe, Y. (1994). *Post-disaster residential rebuilding: a study for California governor's Office of Emergency Services (OES).* (Working paper; Vol. 608). University of California.

CPB. (2013). *Toekomst voor de zorg.* Centraal Planbureau. https://www.cpb.nl/publicatie/toekomst-voor-de-zorg

CPB. (2022). *Ontwikkeling van het arbeidsinkomen tijdens corona.* Centraal Planbureau. https://www.cpb.nl/sites/default/files/omnidownload/CPB-Coronapublicatie-Ontwikkeling-van-het-arbeidsinkomen-tijdens-corona.pdf

CPB. (2023). *Concept Macro Economische Verkenning 2024.* Centraal Planbureau. https://www.cpb.nl/sites/default/files/omnidownload/CPB-Raming-Concept-Macro-Economische-Verkenning-2024-augustus-2023.pdf

De Bruijn, K., & Slager, K. (2022). *Wat als ' de waterbom' elders in Nederland was gevallen?* Deltares. https://publications.deltares.nl/11206890_010_0006.pdf

De Gelderlander. (1995, januari 30). Evacuatie op gang in Middelaar; Er mag geen auto het dorp meer in. *De Gelderlander.*

De Graaf, K. (2022). Naar een dynamische (Ed.) milieuvergunning. Bye-bye bestaande rechten? In *Milieurecht in transitie: Jubileumbundel 40 jaar VMR* (pp. 87–95). Boom Juridisch.

De Koning, K., & Filatova, T. (2020). Repetitive floods intensify outmigration and climate gentrification in coastal cities. *Environmental Research Letters, 15*(3), Article 034008.

De Voogd, J. & Cuperus, R. (2021). *Atlas van afgehaakt Nederland. Over buitenstaanderse en gevestigden.*

De Vries, S., Buijs, A. E., & Snep, R. P. H. (2020). Environmental Justice in The Netherlands: Presence and Quality of Greenspace Differ by Socioeconomic Status of Neighbourhoods. *Sustainability, 12*(15), 5889. https://www.mdpi.com/2071-1050/12/15/5889

de Vries, A., Werner, G., Wijlhuizen, E., Toom, V., Bovens, M., & Hulscher, S. (2024). *Justice in climate policy: Distributing climate costs fairly.* Springer Nature.

Dekker, P., & Van der Meer, T. (2011). Geen vertrouwen meer? In L. C. J. M. Halman & I. J. P. Sieben (Eds.), *Respect man! Tolerantie, solidariteit en andere moderne waarden* (pp. 197–210). Celsus.

Deltares. (2024). *Signaal 3: Kansen en gevaren van klimaatrisicoinformatie op huisniveau.* https://www.deltaprogramma.nl/documenten/publicaties/2024/11/29/signaal-3.-kansen-en-gevaren-van-klimaatinformatie-op-huisniveau

Denters, B., & Klok, P.-J. (2010). Rebuilding Roombeek: Patterns of Citizen Participation in Urban Governance. *Urban Affairs Review, 45*(5), 583–607. https://doi.org/10.1177/1078087409356756

Derkzen, M. L., Van Teeffelen, A. J. A., & Verburg, P. H. (2017). Green infrastructure for urban climate adaptation: How do residents' views on climate impacts and green infrastructure shape adaptation preferences? *Landscape and Urban Planning, 157*, 106–130. https://doi.org/10.1016/j.landurbplan.2016.05.027

Diakakis, M., Skordoulis, M., & Savvidou, E. (2021). The relationships between public risk perceptions of climate change, environmental sensitivity and experience of extreme weather-related disasters: Evidence from Greece. *Water, 13*(20), 2842.

Dijkstra, A. (2012). *Sociale opbrengsten van onderwijs.* Vossiuspers UvA. https://pure.uva.nl/ws/files/1615318/122688_PDF_7303Dijkstra_webversie.pdf

Ditlevsen, P., & Ditlevsen, S. (2023). Warning of a forthcoming collapse of the Atlantic meridional overturning circulation. *Nature Communications, 14*(1), 1–12.

Don, C. (2025). *Klimaatgetto's: hoe klimaatverandering onze huizen binnendringt.* Walburg Pers.

Dryhurst, S., Schneider, C. R., Kerr, J., Freeman, A. L., Recchia, G., Van Der Bles, A. M., Spiegelhalter, D., & Van Der Linden, S. (2022). Risk perceptions of COVID-19 around the world. In J. K. Wardman & R. Löfstedt (Eds.), *COVID-19* (pp. 162–174). Routledge.

Dückers M. (2023). De rampen die we samen creëren: Terugkerende patronen in crisisaanpak. *De Psycholoog* 60(5). https://www.tijdschriftdepsycholoog.nl/wetenschap/de-rampen-die-we-samen-creeren/

Duijndam, S. J., Botzen, W. J. W., Endendijk, T., de Moel, H., Slager, K., & Aerts, J. C. J. H. (2023). A look into our future under climate change? Adaptation and migration intentions following extreme flooding in the Netherlands. *International journal of disaster risk reduction, 95*, 103840. https://doi.org/10.1016/j.ijdrr.2023.103840

Dynes, R. (2003). Finding order in disorder: Continuities in the 9–11 response. *International Journal of Mass Emergencies & Disasters, 21*(3), 9–23.

Ebi, K. L., Vanos, J., Baldwin, J. W., Bell, J. E., Hondula, D. M., Errett, N. A., Hayes, K., Reid, C. E., Saha, S., Spector, J., & Berry, P. (2021). Extreme weather and climate change: Population health and health system implications. *Annual Review of Public Health, 42*, 293–315. https://doi.org/10.1146/annurev-publhealth-012420-105026

EEA. (2024). *European Climate Risk Assessment.* https://www.eea.europa.eu/en/analysis/publications/european-climate-risk-assessment#:~:text=The%20first%20European%20Climate%20Risk%20Assessment%20%28EUCRA%29%20identifies,infrastructure%2C%20water%20resources%2C%20financial%20stability%2C%20and%20people%E2%80%99s%20health

Eimers, D. (2020, 22 April). Een tweede Groningen. *De Groene Amsterdammer.* https://www.groene.nl/artikel/een-tweede-groningen

Ember, C. R., Skoggard, I., Felzer, B., Pitek, E., & Jiang, M. (2021). Climate variability, drought, and the belief that high gods are associated with weather in nonindustrial societies. *Weather, Climate, and Society, 13*(2), 259–272.

Engbersen, G. (2022). De maatschappelijke impact van COVID-19—Sociologie van een pandemie. *De Sociologie en de Pandemie: Inzichten en Vooruitblik na Twee Jaar Coronacrisis.* https://pure.uvt.nl/ws/files/65910933/Sociologie_en_de_pandemie_compleet_2022_pdf

Engbersen, G., Van Bochove, M., de Boom, J., Bussemaker, J., El Farisi, B., Krouwel, A., Van Lindert, J., Rusinovic, K., Snel, E., & Van Heck, L. (2021). *De laag-vertrouwensamenleving. De maatschappelijke impact van COVID-19 in Amsterdam, Den Haag, Rotterdam & Nederland, vijfde meting.* Kenniswerkplaats Leefbare Wijken.

Erikson, K. (1995). *A new species of trouble: The human experience of modern disasters.* Norton & Company Inc.

Etkin, D. A., Mamuji, A. A., & Clarke, L. (2018). Disaster risk analysis part 1: The importance of including rare events. *Journal of Homeland Security and Emergency Management, 15*(2), 20170007.

Eurostat. (2024). *Price of olive oil up 50% in one year.* https://ec.europa.eu/eurostat/web/products-eurostat-news/w/ddn-20240227-1

Flanagan, B. E., Gregory, E. W., Hallisey, E. J., Heitgerd, J. L., & Lewis, B. (2011). A social vulnerability index for disaster management. *Journal of Homeland Security and Emergency Management, 8*(1), 0000102202154773551792.

Forgrave, R., Elliott, E. M., & Bain, D. J. (2022). Event scale hydrograph responses highlight impacts of widespread stream burial and urban infrastructure failures. *Hydrological Processes, 36*(5), Article e14584.

Forrest, S. A., Trell, E. -M., & Woltjer, J. (2020). Socio-spatial inequalities in flood resilience: Rainfall flooding in the city of Arnhem. *Cities, 105*, 102843. https://doi.org/10.1016/j.cities.2020.102843

Fothergill, A., & Peek, L. A. (2004). Poverty and disasters in the United States: A review of recent sociological findings. *Natural Hazards, 32*(1), 89–110. https://doi.org/10.1023/B:NHAZ.0000026792.76181.d9

Frank, T. (2020). Flooding disproportionately harms black neighborhoods. *Scientific American.* https://www.scientificamerican.com/article/flooding-disproportionately-harms-black-neighborhoods/

Frederik, J. (2024). Hittedoden door klimaatverandering, wat weten we nu écht? *De Correspondent.* https://decorrespondent.nl/15521/hittedoden-door-klimaatverandering-wat-weten-we-nu-echt/7d8d0ef8-92ba-0e38-0d1c-3bc343bbda9e

Municipality of Amsterdam / Gemeente Amsterdam. (2024). *Op de vlucht voor poepvliegen en cholera: de Amsterdamse buitenplaatsen van weleer.* https://www.amsterdam.nl/nieuws/achtergrond/buitenplaatsen/

Gemeente Amsterdam. (n.d.). *Klimaatadaptatie.* https://maps.amsterdam.nl/klimaatadaptatie/

Municipality of The Hague / Gemeente Den Haag. (2021). *Visie herinrichting Bomenbuurt.* https://hoplrcontent.blob.core.windows.net/attachments/Visie%20herinrichting%20bomenbuurt_343116eebb0de0-0bb9-4938-85c6-f92e21646647.pdf

Gemeente Den Haag. (n.d.a). *Den Haag in cijfers.* https://denhaag.incijfers.nl/home?ReturnUrl=%2fdashboard%2fnl-nl%2fsegregatiemonitor%2f3—wonen-en-leefomgeving

Gemeente Den Haag. (n.d.b). *Leeftijd, achtergrond en sociaal-economische status.* https://denhaag.incijfers.nl/dashboard/nl-nl/segregatiemonitor/2—leeftijd—achtergrond-en-sociaal-economische-status

Gemeente Heemstede. (n.d.). *Klimaatkaarten.* https://www.heemstede.nl/klimaatkaart

Gemeente Rotterdam. (2023). *Rotterdam gaat voor groen: de groenagenda 2023–2026* (23bb004220). https://gemeenteraad.rotterdam.nl/Agenda/Document/1d7e338f-56fd-4c1e-8246-deb5abbe8fe9?documentId=df6b2db4-9d28-468a-9ca4-56e579e3889d&agendaItemId=3a7c6852-5ea1-4e7c-9064-13ba53a0f8dd

Gemeente Valkenburg. (2021). *Noodverordening.* https://www.valkenburg.nl/data/downloadables/5/0/2/noodverordening-21-juli-2021.pdf

Geng, T., Jia, F., Cai, W., Wu, L., Gan, B., Jing, Z., Li, S., & McPhaden, M. J. (2023). Increased occurrences of consecutive La Niña events under global warming. *Nature, 619*(7971), 774–781.

Gerretschen, J. (2022). Klimaatadaptatie in het vaarwater van sociale gelijkheid. *Water Governance Tijdschrift.* https://www.stowa.nl/sites/default/files/assets/PROJECTEN/Projecten%202022/Water%20Governance/Edities/WAGO_2022-02_Klimaatactie_DEF.pdf

Gerson, S. (2024). Affiliation vs. Alienation. *Psychoanalytic Dialogues, 34*(3), 280–287.

Gersons, B. (2024). Betoog: Voorkom op tijd de ramp na de ramp. *VNG Magazine, 3.* https://vng.nl/artikelen/betoog-voorkom-op-tijd-de-ramp-na-de-ramp

Gijsberts, M. et al. (2022). Vaccinatiebereidheid en vertrouwen in instituties onder Nederlanders met en zonder migratieachtergrond. https://stukroodvlees.nl/vaccinatiebereidheid-en-vertrouwen-in-instituties-onder-nederlanders-met-en-zonder-migratieachtergrond/

Gladwin, H., & Peacock, W. G. (1997). Warning and evacuation: A night for hard houses. In W. G. Peacock, & H. Gladwin (Eds.), *Hurricane Andrew. Ethnicity, Gender and the Sociology of Disasters* (pp. 52–74). Routledge.

Glas, I., & Mulder, L. (2018). Diversiteit naar herkomst en sociale cohesie. *De nieuwe verscheidenheid, toenemende diversiteit naar herkomst in Nederland.* In: R. Jennissen, G. Engbersen, M. Bokhorst & M. Bovens. WRR Verkenning nr 38.

Greene, G., Paranjothy, S., & Palmer, S. R. (2015). Resilience and vulnerability to the psychological harm from flooding: The role of social Cohesion. *American Journal of Public Health, 105*(9), 1792–1795. https://doi.org/10.2105/ajph.2015.302709

Grisham, E. L., Jones, N. M., Silver, R. C., & Holman, E. A. (2023). Do Past events sow future fears? Temporal disintegration, distress, and fear of the future following collective Trauma. *Clinical Psychological Science, 11*(6), 1064–1074.

Green–Blue Schoolyards / Groenblauwe Revolutieschoolpleinen. (n.d.). *Schoolpleinen.* https://groenblauwerevolutieschoolpleinen.nl

Green–Blue Revolution Schoolyards / Groenblauwe Schoolpleinen. (n.d.). *Samen maken we onze schoolpleinen duurzaam.* https://www.groenblauweschoolpleinen.nl

Groeneveld, G., Kodde, J., Waning, S. (2024, juli 21). Extreme wateroverlast in Twente: drijvende auto's in Enschede, A1 tussen Oldenzaal en Duitse grens urenlang afgesloten. *Tubantia.* https://www.tubantia.nl/enschede/extreme-wateroverlast-in-twen-te-drijvende-autos-in-enschede-a1-tussen-oldenzaal-en-duitse-grens-urenlang-afgesloten~a616b050/?referrer=https%3A%2F%2Fwww.google.com%2F

Guy, S., Kashima, Y., Walker, I., & O'Neill, S. (2014). Investigating the effects of knowledge and ideology on climate change beliefs. *European Journal of Social Psychology, 44*(5), 421–429. https://doi.org/10.1002/ejsp.2039

Hallegatte, S., & Rozenberg, J. (2017). Climate change through a poverty lens. *Nature Climate Change, 7*(4), 250–256. https://doi.org/10.1038/nclimate3253

Hallin, C., Martinez, G., Hofstede, J. L., Jensen, J., Baron, N., Heimann, T., Kroon, A., Arns, A., Almström, B., & Sørensen, P. (2021). A comparative study of the effects of the 1872 storm and coastal flood risk management in Denmark, Germany, and Sweden. *Water, 13*(12), 1697.

't Hart, P. t., Pot, W., & Biesbroek, G. (2023). Klimaatadaptatie: hoe leggen we onze oogkleppen af? In T. Overmans, M. Honingh, & M. Noordegraaf (Eds.), *Maatschappelijke bestuurskunde: Hoe verbindende bestuurskundigen (kunnen) inspelen op maatschappelijke vraagstukken* (pp. 29–50). Boom.

Heimann, T., & Mallick, B. (2016). Understanding Climate Adaptation Cultures in Global Context: Proposal for an Explanatory Framework. *Climate, 4*(4), 59. https://www.mdpi.com/2225-1154/4/4/59

Helsloot, I., Scholtens, A. (2015). *Krachten rond de risico-regelreflex beschreven en geïllustreerd in 27 voorbeelden.* Boom Juridische Uitgevers.

Hettige, S. (2023). Introduction: Sociology of Disasters. In A. Singh (Ed.), *International Handbook of Disaster Research* (pp. 1529–1534). Springer.

Hiddink, J. (2023). *Ruim helft van akkerbouwers kampt met opbrengstderving door weersextremen.* Nieuwe Oogst. https://www.nieuweoogst.nl/nieuws/2024/05/03/ruim-helft-van-akkerbouwers-kampt-met-opbrengstder-ving-door-weersextremen

Hore, K., Kelman, I., Mercer, J., Gaillard, J. (2018). Climate Change and Disasters. In H. Rodríguez, W. Donner, & J. Trainor (Eds.), *Handbook of Disaster Research.* Springer. https://doi.org/10.1007/978-3-319-63254-4_8

Huisje Boompje Beter. (n.d.). *Blauwe tuinen zijn 'toffe tuinen'.* https://www.huisjeboompjebeter.nl/tilburg/blauwe-tuinen/

HWBP. (2021). *Stappenplan beoordeling sober en doelmatig ontwerp.* Hoogwaterbeschermingsprogramma. https://www.hwbp.nl/kennisbank/sober-en-doelmatig/documenten/handreikingen/2021/10/25/stappenplan-sober-en-doelmatig

I&O Research (2023). *Risico- en crisisbarometer voorjaar 2023* (2023/092). Nationaal Coördinator Terrorismebestrijding en Veiligheid (NCTV). https://www.nctv.nl/documenten/rapporten/2023/06/05/risico--en-crisisbarometer---voorjaar-2023

IenM. (2016). *Nationale klimaatadaptatiestrategie (NAS).* Ministerie van Infrastructuur en Milieu. https://klimaatadaptatienederland.nl/publish/pages/120542/nas_rapport_5_2-1-1.pdf

IenW. (2023). *Nationaal Uitvoeringsprogramma Klimaatadaptatie Slimmer, intensiever, voor en door iedereen.* Ministerie van Infrastructuur en Waterstaat. https://www.rijksoverheid.nl/documenten/rapporten/2023/11/17/bijlage-2-nationaal-uitvoeringsprogramma-klimaatadaptatie-nup-ka

IenW. (2024). *Brief Water en Bodem naar aanleiding van het Tweeminutendebat Water 8 oktober 2024* [Kamerbrief]. Ministerie van Infrastructuur en Waterstaat. https://open.overheid.nl/documenten/dpc-d278a56998fb4253af9f35329e230b7bacafc217/pdf

Informatiepunt Leefomgeving. (n.d.). *Landelijke Coördinatiecommissies bij watercrises.* https://iplo.nl/thema/water/crisismanagement-water/landelijke-coordinatiecommissies/

Inspectie JenV. (2018). *Onderzoek naar het systeem van de rampenbestrijding op de BES-Eilanden.* Ministerie van Justitie en Veiligheid. https://www.inspectie-jenv.nl/documenten/2018/06/27/reactie-minister-op-onderzoek-naar-het-systeem-van-de-rampenbestrijding-op-de-bes-eilanden

Inspectorate of Justice and Security / Inspectie JenV (2018: 58-59). Onderzoek naar het systeem van derampenbestrijding op de BES-Eilanden. Ministerie van Justitie en Veiligheid. https://www.inspectie-jenv.nl/documenten/2018/06/27/reactie-minister-op-onderzoek-naar-het-systeem-van-de-rampenbestrijding-op-de-bes-eilanden

Inspectie van het Onderwijs. (2022a). *De Staat van het Onderwijs 2022.* https://www.onderwijsinspectie.nl/documenten/rapporten/2022/04/13/de-staat-van-het-onderwijs-2022

Inspectie van het Onderwijs. (2022b). *Technisch rapport huisvesting.* https://www.onderwijsinspectie.nl/documenten/rapporten/2022/04/13/technisch-rapport-huisvesting-svho-2022

Investico. (2024). *Onderzoeksverantwoording Hittestress.* https://www.platform-investico.nl/media/pages/onderzoeken/2-miljoen-ouderen-in-woning-met-risico-op-oververhitting/4b2cd50847-1720512693/onderzoeksverantwoording-hittestress.pdf

IPCC. (2014). *Summary for policymakers* (Climate Change 2014: Impacts, Adaptation and Vulnerability. Part A: Global and Sectoral Aspects. Contribution of Working Group II to the Fifth Assessment Report of the Intergovernmental Panel on Climate Change). C. U. Press.

IPCC. (2022). *Climate Change 2022: Impacts, Adaptation and Vulnerability. Working Group II Contribution to the Sixth Assessment Report of the Intergovernmental Panel on Climate Change.* C. U. Press. https://www.ipcc.ch/report/ar6/wg2/

IPCC. (2023). *Climate Change 2023. Synthesis Report. Summary for Policymakers.* https://www.ipcc.ch/report/ar6/syr/downloads/report/IPCC_AR6_SYR_SPM.pdf

Ireland, P., & Thomalla, F. (2011). The role of collective action in enhancing communities' adaptive capacity to environmental risk: an exploration of two case studies from Asia. *PLoS Curr, 3,* Rrn1279. https://pubmed.ncbi.nlm.nih.gov/22045442/

Islam, N. W., & Winkel, J. (2017). *Climate change and social inequality* (DESA Working Paper, No. 152). Verenigde Naties, Department of Economic & Social Affairs. https://bhekisisa.org/wp-content/uploads/2023/09/wp152_2017.pdf

Jensen, L. (2020). De strijd tegen het water in woord en beeld. Verbindingen tussen heden en verleden. In L. Jensen, & A. Duiveman (Eds.), *Welke verhalen vertellen we? Narratieve strategieën rondom water- beheer en zeespiegelstijging* (pp. 6–13). Radboud Universiteit. https://repository.ubn.ru.nl/handle/2066/227293

Jensen, L. (2024). *Rampen: Een nieuwe geschiedenis van Nederland.* Uitgeverij Prometheus.

Jo, H.-S., & Ham, Y.-G. (2023). Enhanced joint impact of western hemispheric precursors increases extreme El Niño frequency under greenhouse warming. *Nature Communications, 14*(1), 6356.

Kanne, P., & Driessen, M. (2021). *Vertrouwen in overheid na drie grote kwesties* (2021/068). I&O Research.

Kasmalkar, I., Wagenaar, D., Bill-Weilandt, A., Choong, J., Manimaran, S., Lim, T. N., Rabonza, M., & Lallemant, D. (2024). Flow-tub model: A modified bathtub flood model with hydraulic connectivity and path-based attenuation. *MethodsX, 12,* Article 102524.

Kasperson, R. E., Renn, O., Slovic, P., Brown, H. S., Emel, J., Goble, R., Kasperson, J. X., & Ratick, S. (1988). The social amplification of risk: A conceptual framework. *Risk Analysis, 8*(2), 177–187. https://doi.org/10.1111/j.1539-6924.1988.tb01168.x

KCAF. (n.d.a). *Landelijk loket funderingsproblematiek*. Kennis Centrum Aanpak Funderingsproblematiek. https://www.kcaf.nl/landelijk-loket-funderingsproblematiek/

KCAF. (n.d.b). *De Nationale Funderingsramp (deel 2): Funderingsherstel vraagt een lange adem*. Kennis Centrum Aanpak Funderingsproblematiek. https://www.kcaf.nl/de-nationale-funderingsramp-deel-2/

Keenan, J. M., Hill, T., & Gumber, A. (2018). Climate gentrification: From theory to empiricism in Miami-Dade County. *Florida. Environmental Research Letters, 13*(5), Article 054001.

Keller, R. C. (2015). *Fatal Isolation*. University of Chicago Press.

Kind, J., Koers, G., de Nijs, T., Koopmans, R., van Popering, J., Visser, V., & Damen, M. (2020). *Verkenning naar de wisselwerking tussen sociale veerkracht en klimaatadaptatie*. Deltares.

Kitt, S., Axsen, J., Long, Z., & Rhodes, E. (2021). The role of trust in citizen acceptance of climate policy: Comparing perceptions of government competence, integrity and value similarity. *Ecological Economics, 183*, Article 106958.

Kleinhans, R. (2017). False promises of co-production in neighbourhood regeneration: The case of Dutch community enterprises. *Public Management Review, 19*(10), 1500–1518. https://doi.org/10.1080/14719037.2017.1287941

Klimaat voor Ruimte. (2006). *Naar een klimaatbestendig Nederland*. https://edepot.wur.nl/254919

Klimaatadaptatie Nederland. (2020). *Handreiking Slim Koppelen klimaat-adaptatie voor gemeenten*. https://klimaatadaptatienederland.nl/hulpmiddelen/overzicht/handreiking-slim-koppelen/

Klimaatadaptatie Nederland. (n.d.a). *Impulsregeling klimaatadaptatie*. https://klimaatadaptatienederland.nl/beleid/nationale-aanpak/dpra/deltaplan/stimuleren-faciliteren/impulsregeling-klimaatadaptatie/

Klimaatadaptatie Nederland. (n.d.b). *Voorbeeldlijst maatregelen Impulsregeling klimaatadaptatie*. https://klimaatadaptatienederland.nl/publish/pages/179576/voorbeeldlijst_maatregelen_impulsregeling.pdf

Klimaateffectatlas. (n.d.). *Klimaateffectatlas*. https://www.klimaateffectatlas.nl/nl/

Klinenberg, E. (2002). *Heat wave: A social autopsy of disaster in Chicago*. University of Chicago Press.

Klinenberg, E. (2015). *Heat wave: A social autopsy of disaster in Chicago (2nd edition)*. University of Chicago Press.

Klinenberg, E. (2018). *Palaces for the people: How social infrastructure can help fight inequality, polarization, and the decline of civic life*. Crown.

KNMI. (2023). *KNMI'23-klimaatscenario's*. Koninklijk Nederlands Meteorologisch Instituut. https://www.knmi.nl/kennis-en-datacentrum/achtergrond/knmi-23-klimaatscenario-s

KNMI. (2025). *De staat van ons klimaat 2024*. Koninklijk Nederlands Meteorologisch Instituut. https://cdn.knmi.nl/system/data_center_publications/files/000/072/295/original/KNMI-publicatie_25-01.pdf

Kolner, C., & Engbersen, G. (2024). Vitale sociale infrastructuur bereikt burgers en redt levens. *Sociale Vraagstukken*. https://www.socialevraagstukken.nl/vitale-sociale-infrastructuur-bereikt-burgers-en-redt-levens/

Korsten, A. F. A., & de Goede, P. (2006). De betekenis van sociaal kapitaal voor vertrouwen in overheidsbestuur. In A. F. A. Korsten & P. de Goede (Eds.), *Bouwen aan vertrouwen in het openbaar bestuur* (pp. 175–191). Elsevier.

Latham, A., & Layton, J. (2019). Social infrastructure and the public life of cities: Studying urban sociality and public spaces. *Geography Compass, 13*(7), Article e12444.

Lee, T. M., Markowitz, E. M., Howe, P. D., Ko, C.-Y., & Leiserowitz, A. A. (2015). Predictors of public climate change awareness and risk perception around the world. *Nature Climate Change, 5*(11), 1014–1020.

Leichenko, R., & Silva, J. A. (2014). Climate change and poverty: vulnerability, impacts, and alleviation strategies. *WIREs Climate Change*, *5*(4), 539–556. https://doi.org/10.1002/wcc.287

Lenton, T. M., Held, H., Kriegler, E., Hall, J. W., Lucht, W., Rahmstorf, S., & Schellnhuber, H. J. (2008). Tipping elements in the Earth's climate system. *Proceedings of the National Academy of Sciences, 105*(6), 1786–1793.

Lenton, T. M., Myerscough, R. J., Marsh, R., Livina, V. N., Price, A. R., Cox, S. J., & Team, G. (2009). Using GENIE to study a tipping point in the climate system. *Philosophical Transactions of the Royal Society A: Mathematical, Physical and Engineering Sciences*, *367*(1890), 871–884.

Levac, J., Toal-Sullivan, D., & O'Sullivan, T. L. (2012). Household Emergency Preparedness: A Literature Review. *Journal of Community Health, 37*(3), 725–733. https://doi.org/10.1007/s10900-011-9488-x

Ligtvoet, W., & Knoop, J. (2014). *Kleine kansen—grote gevolgen: Slachtoffers en maatschappelijke ontwrichting als focus voor waterveiligheidsbeleid*. https://www.pbl.nl/publicaties/kleine-kansen-grote-gevolgen

Limaheluw, J., Schets, C., Vermeulen-Henstra, L., De Roda & Husman, A. (2023). Klimaatverandering en wateroverdraagbare infectieziekten in Nederland. *Nederlands Tijdschrift voor Medische Microbiologie, 31*(2). https://www.nvmm.nl/media/5259/21714-054.pdf

Liu, J., Zhao, M., & Wang, Y. (2020). Impacts of government subsidies and environmental regulations on green process innovation: A nonlinear approach, *Technology in Society*, 63, 101417, ISSN 0160-791X. https://doi.org/10.1016/j.techsoc.2020.101417

LNV. (2020). *Actieprogramma klimaatadaptatie landbouw*. Ministerie van Landbouw, Natuur en Voedselkwaliteit. https://open.overheid.nl/documenten/ronl-35db3cf1-e9e6-4663-8963-4844da9ce1d1/pdf

Lob van Gennep. (2022). *Voorkeursbeslissing Lob van Gennep*. https://www.lobvangennep.nl/applications/lobvangennep/files/Downloads/voorkeursbeslissing/Voorkeursbeslissing-Lob-van-Gennep.pdf

Lynn, B. H., Healy, R., & Druyan, L. M. (2009). Investigation of Hurricane Katrina characteristics for future, warmer climates. *Climate Research, 39*(1), 75–86.

Maas, J., Verheij, R. A., de Vries, S., Spreeuwenberg, P., Schellevis, F. G., & Groenewegen, P. P. (2009). Morbidity is related to a green living environment. *Journal of Epidemiology and Community Health, 63*(12), 967–973. https://doi.org/10.1136/jech.2008.079038

Mackay, C. M., Schmitt, M. T., Lutz, A. E., & Mendel, J. (2021). Recent developments in the social identity approach to the psychology of climate change. *Current Opinion in Psychology, 42*, 95–101.

Marikyan, D., & Papagiannidis, S. (2023). Protection motivation theory: A review. In S. Papagiannidis (Ed.), *TheoryHub Book: This handbook is based on the online theory resource: TheoryHub* (pp. 78–93). https://open.ncl.ac.uk/theory-library/TheoryHubBook.pdf

Martinez, G., Costas, S., & Ferreira, Ó. (2020). The role of culture for coastal disaster risk reduction measures: Empirical evidence from northern and southern Europe. *Advances in Climate Change Research, 11*(4), 297–309.

Matthews, V., Longman, J., Bennett-Levy, J., Braddon, M., Passey, M., Bailie, R. S., & Berry, H. L. (2020). Belonging and inclusivity make a resilient future for all: A cross-sectional analysis of post-flood social capital in a diverse Australian rural community. *International Journal of Environmental Research and Public Health, 17*(20), 7676.

Mecanoo Architecten. (2023). *Maasterras Stedenbouwkundig Masterplan*.

Meijer, F. (2022). *Verbonden door rampspoed: Rampen en natievorming in negentiende-eeuws Nederland*. Uitgeverij Verloren.

Meyer, M. A. (2018). Social Capital in Disaster Research. In H. Rodríguez, W. Donner, & J. E. Trainor (Eds.), *Handbook of Disaster Research* (pp. 263–286). Springer International Publishing. https://doi.org/10.1007/978-3-319-63254-4_14

Meyer, R., & Kunreuther, H. (2017). *The ostrich paradox: Why we underprepare for disasters*. University of Pennsylvania Press.

Miles-Novelo, A., & Anderson, C. A. (2019). Climate change and psychology: Effects of rapid global warming on violence and aggression. *Current Climate Change Reports, 5*(1), 36–46. https://doi.org/10.1007/s40641-019-00121-2

Morss, R. E., Cuite, C. L., & Demuth, J. L. (2024). What predicts hurricane evacuation decisions? The importance of efficacy beliefs, risk perceptions, and other factors. NPJ *Natural Hazards, 1*(1), 24. https://doi.org/10.1038/s44304-024-00025-8

Movisie, & Verweij-Jonker. (2024). *De sociale basis versterken, investeren in de toekomst. Verandertheorie voor overheden en maatschappelijke partners.* https://www.movisie.nl/publicatie/sociale-basis-versterken-investeren-toekomst

Mulder, A. C., Pijnacker, R., de Man, H., van de Kassteele, J., van Pelt, W., Mughini-Gras, L., & Franz, E. (2019). "Sickenin' in the rain"—increased risk of gastrointestinal and respiratory infections after urban pluvial flooding in a population-based cross-sectional study in the Netherlands. *BMC Infectious Diseases, 19*(1), 377. https://doi.org/10.1186/s12879-019-3984-5

Munich Re. (2024). *Record thunderstorm losses and deadly earthquakes: the natural disasters of 2023.* https://www.munichre.com/en/company/media-relations/media-information-and-corporate-news/media-information/2024/natural-disaster-figures-2023.html#video

Nakai, F., & Nakano, G. (2023). Community-mediated individual disaster preparedness practices: A case study in Kochi, Japan. *International Journal of Disaster Risk Reduction, 86*, Article 103532.

Nasi, L., Jans, V., & Steg, L. (2024). Individual transilience in the face of the covid-19 pandemic. *Journal of Environmental Psychology, 93*, Article 102188.

Nationaal Deltaprogramma. (2022). *Advies Signaalgroep Deltaprogramma 2022.* https://www.deltaprogramma.nl/documenten/publicaties/2022/11/10/adviesbrief-signaalgroep-2022

Nationaal Deltaprogramma. (2024a). *Advies Signaalgroep Deltaprogramma 2024.* https://www.deltaprogramma.nl/documenten/publicaties/2024/11/29/advies-signaalgroep-deltaprogramma-2024

Nationaal Deltaprogramma. (2024b). *Nu voor later. Deltafonds* (hoofdstuk 07). https://minienw.sitearchief.nl/?subsite=dp2024deltaprogramma#archive

Nationaal Deltaprogramma. (2024c). *Naar een nieuwe balans in de leefomgeving: Ruimte voor leven met water*. https://dp2025.deltaprogramma.nl

Nationaal Deltaprogramma. (n.d.). *Deltabeslissing Waterveiligheid.* https://www.deltaprogramma.nl/themas/waterveiligheid/deltabeslissing

Dutch National Disaster Relief Found / Nationaal Rampenfonds. (n.d.). *Het Nationaal Rampenfonds.* https://nationaalrampenfonds.nl/

Netherlands Court of Audit / Nationale ombudsman. (2023). *Herstel bieden: een vak apart* (2023/157). https://www.nationaleombudsman.nl/system/files/bijlage/Herstel%20bieden_een%20vak%20apart.pdf

NCTV. (2022). *Themarapportage klimaat en natuurrampen 2022.* Nationaal Coördinator Terrorismebestrijding en Veiligheid. https://www.nctv.nl/documenten/publicaties/2022/09/26/themarapportage-klimaat-en-natuurrampen-2022

NCTV. (2024). *Handreiking Beter voorbereid op rampen en crises.* Nationaal Coördinator Terrorismebestrijding en Veiligheid. https://www.nctv.nl/documenten/publicaties/2024/05/23/handreiking-beter-voorbereid-op-rampen-en-crises

National Mortgage Guarantee / Nationale Hypotheek Garantie (NHG). (2024). *De Woonlastenmonitor*. Nationale Hypotheek Garantie. https://www.nhg.nl/woonlastenmonitor/

Nickerson, R. S. (1998). Confirmation Bias: A Ubiquitous Phenomenon in Many Guises. *Review of General Psychology, 2*(2), 175–220. https://doi.org/10.1037/1089-2680.2.2.175

NIPV. (2019). *Maatschappelijk herstel na grootschalige overstromingen.* Nationaal Instituut Publieke Veiligheid. https://nipv.nl/wp-content/uploads/2022/02/20191017-IFV-Wave2020-Maatschappelijk-herstel-na-grootschalige-overstromingen.pdf

NIPV. (2023). *Natuurbrandsignaal '23.* Nationaal Instituut Publieke Veiligheid. https://wp-content/uploads/2023/04/20230123-NIPV-Natuurbrandsignaal-23.pdf

NIPV. (2024). *Versterken van veerkracht. Naar een gezamenlijke aanpak van ongekende crises.* Nationaal Instituut Publieke Veiligheid. https://nipv.nl/wp-content/uploads/2024/05/20240501-NIPV-Crisisplan-Versterken-van-veerkracht.pdf

NJI. (2023). *Mentaal welbevinden van de jeugd: lessen uit de coronacrisis.* Nationaal Jeugdinstituut. https://www.nji.nl/publicaties/mentaal-welbevinden-van-de-jeugd-lessen-uit-de-coronacrisis

NOS. (2021, July 15). *Kabinet noemt overstromingen Zuid-Limburg formeel een ramp, overheid betaalt deel schade.* Nederlandse Omroep Stichting. https://nos.nl/collectie/13869/artikel/2389502-kabinet-noemt-overstromingen-zuid-limburg-formeel-een-ramp-overheid-betaalt-deel-schade

NOS. (2022, November 29). *Nederlanders niet goed voorbereid op ramp: 'We zijn een kwetsbaar land'.* Nederlandse Omroep Stichting. https://nos.nl/artikel/2454384-nederlanders-niet-goed-voorbereid-op-ramp-we-zijn-een-kwetsbaar-land

NOS. (2024a, May 3). *Bijna twee derde akkerbouwers zegt minder te verdienen door weersextremen.* Nederlandse Omroep Stichting. https://nos.nl/artikel/2519067-bijna-twee-derde-akkerbouwers-zegt-minder-te-verdienen-door-weersextremen

NOS. (2024b, August 20). *Extreem weer steeds groter probleem voor Nederlandse appels en peren.* Nederlandse Omroep Stichting. https://nos.nl/artikel/2533888-extreem-weer-steeds-groter-probleem-voor-nederlandse-appels-en-peren

NOS. (2024c, August 12). *Niet kunnen slapen door de hitte: 'Koelte-ongelijkheid steeds groter probleem'.* Nederlandse Omroep Stichting. https://nos.nl/artikel/2533041-niet-kunnen-slapen-door-de-hitte-koelte-ongelijkheid-steeds-groter-probleem

NOS. (2024d, November 3). *Spaanse koning bekogeld met modder, uitgejouwd bij bezoek aan rampgebied.* Nederlandse Omroep Stichting. https://nos.nl/artikel/2543066-spaanse-koning-bekogeld-met-modder-uitgejouwd-bij-bezoek-aan-rampgebied

NOS. (2024e, March 14). *Vaccinatiegraad onder kinderen in Den Haag nog nooit zo gevaarlijk laag.* Nederlandse Omroep Stichting. https://nos.nl/artikel/2512681-vaccinatiegraad-onder-kinderen-in-den-haag-nog-nooit-zo-gevaarlijk-laag

NPO. (2023). *Langs de IJssel. De doorzetter.* Aflevering 1. Nederlandse Publieke Omroep. https://npo.nl/start/serie/langs-de-ijssel/seizoen-1/de-doorzetter

NRC. (2022, July 22). *Klimaatverschuiving: die lange hete Franse zomers uit de jaren 70 vind je nu in Nederland.* https://www.nrc.nl/nieuws/2022/07/22/klimaatverschuiving-die-lange-hete-franse-zomers-uit-de-jaren-70-vind-je-nu-in-nederland-2-a4137119

NRC. (2024a, 12 January). *De minister gaat het verbieden, maar Arnhem bout nog wél in de uiterwaarden.* https://www.nrc.nl/nieuws/2024/01/12/de-minister-gaat-het-verbieden-maar-arnhem-bouwt-nog-wel-in-de-uiterwaar-den-a4186831

NRC. (2024b, 8 July). *Achtduizend woningen in een van de diepste polders. Kan dat wel veilig?* https://www.nrc.nl/nieuws/2024/07/08/kan-het-dorp-in-die-diepe-polder-wel-veilig-robuust-en-klimaatbestendig-worden-gebouwd-a4859079

Ntontis, E., Drury, J., Amlôt, R., Rubin, G. J., Williams, R., & Saavedra, P. (2020). Collective resilience in the disaster recovery period: Emergent social identity and observed social support are associated with collective efficacy, well-being, and the provision of social support. *British Journal of Social Psychology, 60*(3), 1075–1095. https://doi.org/10.1111/bjso.12434

NVM. (2024). *Analyse woningmarkt 2^e kwartaal 2024.* Nederlandse Vereniging van Makelaars. https://www.nvm.nl/media/lf5f1l52/bijlage-1-analyse-nvm-woningmarkt-2e-kwartaal-2024.pdf

O'Brien, K. (2009). Do values subjectively define the limits to climate change adaptation. *Adapting to Climate Change: Thresholds, Values, Governance, 164*, 180.

O'Sullivan, J. J., Bradford, R. A., Bonaiuto, M., De Dominicis, S., Rotko, P., Aaltonen, J., Waylen, K., & Langan, S. J. (2012). Enhancing flood resilience through improved risk communications. *Natural Hazards and Earth System Sciences, 12*(7), 2271–2282.

Oakley, M., Mohun Himmelweit, S., Leinster, P., & Casado, M. R. (2020). Protection Motivation Theory: A Proposed Theoretical Extension and Moving beyond Rationality—The Case of Flooding. *Water, 12*(7), 1848. https://www.mdpi.com/2073-4441/12/7/1848

OCW. (2023). *Uitvoeringsagenda Klimaat & Energie*. Ministerie van Onderwijs, Cultuur en Wetenschap. https://www.rijksoverheid.nl/documenten/rapporten/2023/10/13/uitvoeringsagenda-klimaat-energie

Oostveen, M. (2023, juni 8). Samen sterk? Burgercollectieven lossen problemen op die de overheid laat liggen. *de Volkskrant*. https://www.volkskrant.nl/kijkverder/v/2023/samen-sterk-burgercollectieven-lossen-problemen-op-die-de-overheid-laat-liggen~v751222/

Ostrom, E., & Ahn, T. -K. (2009). The meaning of social capital and its link to collective action. *Handbook of social capital: The troika of sociology, political science and economics*, 17–35.

Overheid Wettenbank. (2019). *Klimaatwet*. https://wetten.overheid.nl/BWBR0042394/2020-01-01

Palomo-Vélez, G., Perlaviciute, G., Contzen, N., & Steg, L. (2024). Trusting the minister or trusting the mayor? Perceived competence and integrity of central and local Dutch institutions governing energy matters. *Environmental Research Communications, 6*(4), Article 045009.

Park, R. J. (2024). *Slow Burn: The Hidden Costs of a Warming World*. Princeton University Press.

Park, R. J., Goodman, J., Hurwitz, M., & Smith, J. (2020). Heat and learning. *American Economic Journal: Economic Policy, 12*(2), 306–339.

Parliamentary Committee of Inquiry on Natural Gas Extraction in Groningen / Parlementaire enquêtecommissie aardgaswinning Groningen. (2023). *Groningers boven gas*. https://www.tweedekamer.nl/Groningen/rapport

Committee of Inquiry on Childcare Benefits / Parlementaire ondervragingscommissie kinderopvangtoeslag. (2020). *Ongekend onrecht*. https://www.tweedekamer.nl/sites/default/files/atoms/files/20201217_eindverslag_parlementaire_ondervragings-commissie_kinderopvangtoeslag.pdf

PBL. (2014). *Maatschappelijke ontwrichting en overstromingen*. Planbureau voor de Leefomgeving. https://www.pbl.nl/uploads/default/downloads/Maatschappelijke-ontwrichting-en-overstromingen.pdf

PBL. (2015). *Aanpassen aan klimaatverandering – Kwetsbaarheden zien, kansen grijpen* (1454). Planbureau voor de Leefomgeving. https://www.pbl.nl/publicaties/aanpassen-aan-klimaatverandering-kwetsbaarheden-zien-kansen-grijpen

PBL. (2023a). *Nationale klimaatrisicoanalyse 2022–2026* (5044). Planbureau voor de Leefomgeving. https://www.pbl.nl/uploads/default/downloads/pbl-2023-nationale-klimaatrisicoanalyse-2022-2026-uitwerking-methodiek-5044.pdf

PBL. (2023b). *Natuur in en om de stad* (5023). Planbureau voor de Leefomgeving. https://www.pbl.nl/uploads/default/downloads/pbl-2023-natuur-in-en-om-de-stad-5023.pdf

PBL. (2024a). *Klimaatrisico's in Nederland: De huidige stand van zaken* (5359). Planbureau voor de Leefomgeving. https://www.pbl.nl/system/files/document/2024-05/pbl-2024-klimaatrisicos-in-nederland-5359.pdf

PBL. (2024b). *Monitor nationale omgevingsvisie 2024: tweede vervolgmeting* (5313). Planbureau voor de Leefomgeving. https://www.pbl.nl/system/files/document/2024-09/pbl-2024-monitor-nationale-omgevingsvisie-5313.pdf

Peters, L. E. (2021). Beyond disaster vulnerabilities: An empirical investigation of the causal pathways linking conflict to disaster risks. *International Journal of Disaster Risk Reduction, 55*, Article 102092.

Pidgeon, N., & Henwood, K. (2010). The Social Amplification of Risk Framework (SARF): Theory, Critiques, and Policy Implications. In P. Bennett, K. Calman, S. Curtis, & D. Fischbacher-Smith (Eds.), *Risk Communication and Public Health*. Oxford University Press. https://doi.org/10.1093/acprof:oso/9780199562848.003.04

Platform duurzaam Zwolle. (n.d.). *Organisaties*. https://platformduurzaamzwolle.nl/organisaties

Poortinga, W., Demski, C., & Steentjes, K. (2023). Generational differences in climate-related beliefs, risk perceptions and emotions in the UK. *Communications Earth & Environment, 4*(229), 1–8. https://doi.org/10.1038/s43247-023-00870-x

Poumadere, M., Mays, C., Le Mer, S., & Blong, R. (2005). The 2003 heat wave in France: Dangerous climate change here and now. *Risk Analysis: An International Journal, 25*(6), 1483–1494.

Prentice, C. M., Vergunst, F., Minor, K., & Berry, H. L. (2024). Education outcomes in the era of global climate change. *Nature Climate Change, 14*(3), 214–224.

Provincie Gelderland. (2023). *Uitvoeringsagenda klimaatadaptatie 2024–2027*. https://media.gelderland.nl/Uitvoeringsagenda_Klimaatadaptatie_2024_2027_dc8ac817a7.pdf

Provincie Noord-Brabant. (2018). *65 groene daken in de wijk Kattenbosch in Rosmalen*. https://www.klimaatadaptatiebrabant.nl/voorbeelden/voorbeelden-detail/208/65-groene-daken-in-de-wijk-Kattenbosch-in-Rosmalen

Putnam, R. D. (1993). Making democracy work: Civic traditions in modern Italy. *Princeton University Press*. https://doi.org/10.2307/j.ctt7s8r7

Putnam, R. D. (2000). *Bowling alone: The collapse and revival of American community*. Simon & Schuster.

Quarantelli, E. L. (1986). *Planning and management for the prevention and mitigation of natural disasters, especially in a metropolitan context: Initial questions and issues which need to be addressed*. https://udspace.udel.edu/server/api/core/bitstreams/6bd84606-41a0-43a9-a49b-f21a7b678e6a/content

Rackin, H. M., & Weil, F. (2015). *Social capital and the repopulation of New Orleans after Hurricane Katrina*. Annual Meeting of the Population Association of America, San Diego. https://paa2015.populationassociation.org/papers/152366

Renes, R. J. (2021). *De klimaatspagaat: over de psychologische uitdagingen van duurzaam gedrag*. Uitgeverij Creating Tomorrow.

Resilient Delta Initiative. (2023). *Integrated action framework*. Towards a safe and climate-adaptive Maasterras.

Rijksmuseum. (n.d.). IJsvermaak bij een stad. http://hdl.handle.net/10934/RM0001.COLLECT.495302

Government of the Netherlands / Rijksoverheid. (2023). *Beleidstafel wateroverlast en hoogwater*. https://open.overheid.nl/documenten/dpc-327e4ad8c480deec85c2c5b625fa723b8554d0e5/pdf

Government of the Netherlands / Rijksoverheid. (2024). *Nog genoeg ruimte om te bouwen, maar kies verstandige locaties*. https://www.rijksoverheid.nl/actueel/nieuws/2024/04/09/nog-genoeg-ruimte-om-te-bouwen-maar-kies-verstandige-locaties

Government of the Netherlands / Rijksoverheid. (n.d. a). *February 2020: Eerste coronabesmetting in Nederland*. https://www.rijksoverheid.nl/onderwerpen/coronavirus-tijdlijn/februari-2020-eerste-coronabesmetting-in-nederland

Rijkswaterstaat. (n.d. b). *Ruimte voor de rivieren*. https://www.rijkswaterstaat.nl/water/waterbeheer/bescherming-tegen-het-water/maatregelen-om-overstromingen-te-voorkomen/ruimte-voor-de-rivieren

Rijkswaterstaat. (n.d. c) *Waterprojecten*. https://www.rwseconomie.nl/werkwijzers/mkba-bij-mirt-verkenningen/bepalen-effecten-baten/waterprojecten

RIVM. (2012). GGD-*richtlijn medische milieukunde: Gezondheidsrisico's van zomerse omstandigheden*. Rijksinstituut voor Volksgezondheid en Milieu. https://www.rivm.nl/bibliotheek/rapporten/609400007.pdf

RIVM. (2014). *Nationale Risicobeoordeling* 6 (E/609042/14). Rijksinstituut voor Volksgezondheid en Milieu. https://www.rivm.nl/sites/default/files/2019-10/NRB%206%202014.pdf

RIVM. (2021). *Klimaatverandering leidt nu al tot meer sterfte door hitte*. Rijksinstituut voor Volksgezondheid en Milieu. https://www.rivm.nl/nieuws/klimaatverandering-leidt-nu-al-tot-meer-sterfte-door-hitte

RIVM. (2022). *Leidraad risicobeoordeling Rijksbrede Risicoanalyse Nationale Veiligheid*. Rijksinstituut voor Volksgezondheid en Milieu. https://www.rivm.nl/documenten/leidraad-risicobeoordeling-2022

RIVM. (2023). *Gezondheidseconomische aspecten van de covidpandemie*. Rijksinstituut voor Volksgezondheid en Milieu. https://www.rivm.nl/bibliotheek/rapporten/2023-0318.pdf

RIVM. (2024). *Gezondheidseffecten van klimaatverandering: Actualisatie van de huidige klimaatrisico's voor gezondheid*. Rijksinstituut voor Volksgezondheid en Milieu. https://www.rivm.nl/publicaties/gezondheidseffecten-van-klimaat-verandering-actualisatie-van-huidige-klimaatrisicos-voor-gezondheid

RIVM. (n.d.). GGD-*richtlijn medische milieukunde: hitte en gezondheid*. Rijksinstituut voor Volksgezondheid en Milieu. https://www.rivm.nl/ggd-richtlijn-mmk-hitte-gezondheid

Rli. (2024a). *Goed gefundeerd*. Raad voor de leefomgeving en infrastructuur. https://www.rli.nl/publicaties/2024/advies/goed-gefundeerd

Rli. (2024b). *Ruimtelijke ordening in een veranderend klimaat*. Raad voor de leefomgeving en infrastructuur. https://www.rli.nl/sites/default/files/Rli%202024-02%20Klimaatadaptatie%20eindversie.pdf

Rocha, A. D., Vulova, S., Förster, M., Gioli, B., Matthews, B., Helfter, C., Meier, F., Steeneveld, G., Barlow, J., Järvi, L., Chrysoulakis, N., Nicolini, G., & Kleinschmit, B. (2024). Unprivileged groups are less served by green cooling services in major European urban areas. *Nature Cities, 1*(2024), 424–435.

Rogers, R. W. (1975). A Protection Motivation Theory of Fear Appeals and Attitude Change1. *The Journal of Psychology, 91*(1), 93–114. https://doi.org/10.1080/00223980.1975.9915803

Rosenthal, U. (1988). Studies in Holland flood disaster 1953: An essay on the proto-sociology of disaster. *International Journal of Mass Emergencies & Disasters, 6*(3), 233–251.

Rotterdams WeerWoord. (n.d.). *Rotterdams WeerWoord*. Gemeente Rotterdam. https://rotterdamsweerwoord.nl

RVO. (2025). *Zakendoen in Sint Maarten*. Rijksdienst voor Ondernemend Nederland. https://www.rvo.nl/onderwerpen/landen-en-gebieden/sint-maarten

RVS. (2022). *Ruimte maken voor ontmoeting*. Raad voor Volksgezondheid en Samenleving. https://www.raadrvs.nl/adviezen/ruimte-voor-ontmoeting/documenten/publicaties/2022/09/06/ruimte-maken-voor-ontmoeting

Salmons, M., & Woutersen, E. (2024, juli 10). De doorzonwoning als een hitte-eiland: Goed geïsoleerd, maar bloedjeheet. *De Groene Amsterdammer*. https://www.groene.nl/artikel/goed-geisoleerd-maar-bloedjeheet

Sanz-Barbero, B., Linares, C., Vives-Cases, C., González, J. L., López-Ossorio, J. J., & Díaz, J. (2018). Heat wave and the risk of intimate partner violence. *Science of The Total Environment, 644*, 413–419. https://doi.org/10.1016/j.scitotenv.2018.06.368

Saul, J. (2022). *Collective trauma, collective healing: Promoting community resilience in the aftermath of disaster*. Routledge.

Schenk, G. J. (2015). 'Learning from History'?: Chances, problems and limits of learning from historical natural disasters. In *Cultures and Disasters* (pp. 72–87). Routledge.

Schipper, E. L. F. (2020). *Maladaptation: When Adaptation to Climate Change Goes Very Wrong*. One Earth. https://www.cell.com/one-earth/pdf/S2590-3322(20)30483-8.pdf

Schlenker, W., & Roberts, M. J. (2009). Nonlinear temperature effects indicate severe damages to U.S. crop yields under climate change. *Proceedings of the National Academy of Sciences, 106*(37), 15594–15598. https://doi.org/10.1073/pnas.0906865106

Schmeets, H., & Exel, J. (2022). *De Heitjes en het vertrouwen in instituties in Heerlen*. Centraal Bureau voor de Statistiek. https://www.cbs.nl/nl-nl/longread/aanvullende-statistische-diensten/2022/de-heitjes-en-het-vertrouwen-in-instituties-in-heerlen/2-vertrouwen-als-bouwsteen-van-sociale-cohesie

Schmeets, H., Parigini, M., & van Hoof, M. (2016). *Het sociaal kapitaal in Europese regio's*. Centraal Bureau voor de Statistiek. https://www.cbs.nl/nl-nl/achtergrond/2016/43/het-sociaal-kapitaal-in-europese-regio-s

Schoolyard of the Future / Schoolplein van de toekomst. (n.d.). Schoolplein van de toekomst. https://schoolpleinvandetoekomst.nl

Schwaller, N. L., Kelmenson, S., BenDor, T. K., & Spurlock, D. (2020). From abstract futures to concrete experiences: How does political ideology interact with threat perception to affect climate adaptation decisions? *Environmental Science & Policy, 112*, 440–452.

Schwartz, T., Luiten, W., Kok, L. (2020). *Gezondheidsproblemen personen met een lage sociaal-economische status. SEO Economisch Onderzoek.* https://www.seo.nl/wp-content/uploads/2021/01/2021-07-Gezondheidsproblemen-personen-met-een-lage-sociaal-economische-status.pdf

SCP. (2019). *Waar wonen de armen in Nederland?* Sociaal en Cultureel Planbureau. https://digitaal.scp.nl/armoedeinkaart2019/waar-wonen-de-armen-in-nederland/

SCP. (2020). *De sociale staat van Nederland Wonen.* https://digitaal.scp.nl/ssn2020/wonen

SCP. (2023a). *Eigentijdse ongelijkheid: De postindustriële klassenstructuur op basis van vier typen kapitaal.* Sociaal en Cultureel Planbureau. https://www.scp.nl/publicaties/publicaties/2023/03/07/eigentijdse-ongelijkheid

SCP. (2023b). *Samenleving in beweging.* Sociaal en Cultureel Planbureau. https://www.scp.nl/publicaties/publicaties/2023/08/07/samenleving-in-beweging

SCP. (2023c). *Samenleving in beweging.* Sociaal en Cultureel Planbureau. https://www.scp.nl/publicaties/publicaties/2023/08/07/samenleving-in-beweging

SCP. (2024a). *Factsheet Hoe staat het ervoor met de sociale cohesie in Nederland?* Sociaal en Cultureel Planbureau. https://www.scp.nl/publicaties/publicaties/2024/06/04/hoe-staat-ervoor-met-de-sociale-cohesie-in-nederland

SCP. (2024b). *Kennisnotitie Overheidshandelen tijdens en na crises: Vier aandachtspunten voor goed overheidshandelen rond crisissituaties.* Sociaal en Cultureel Planbureau. https://www.scp.nl/publicaties/publicaties/2024/07/18/kennisnotitie-overheidshandelen-tijdens-en-na-crises

SCP. (2024c). *De leefwerelden van arm en rijk.* Sociaal en Cultureel Planbureau. https://www.scp.nl/publicaties/publicaties/2024/04/05/de-leefwerelden-van-arm-en-rijk

SCP. (2025). *Klimaat en Samenleving—Burgerperspectieven.* Sociaal en Cultureel Planbureau. https://www.scp.nl/publicaties/publicaties/2025/04/17/klimaat-en-samenleving

SCP. (n.d.). *Verkenning van toekomstige ontwikkeling van sociale cohesie.* Sociaal en Cultureel Planbureau. https://www.scp.nl/onderzoeksprogramma/verkenning-van-toekomstige-ontwikkeling-van-sociale-cohesie

Slijper, T., Urquhart, J., Poortvliet, P. M., Soriano, B., & Meuwissen, M. P. M. (2022). Exploring how social capital and learning are related to the resilience of Dutch arable farmers. *Agricultural Systems, 198*, 103385. https://doi.org/10.1016/j.agsy.2022.103385

Slovic, P. (1992). *Perception of risk: Reflections on the psychometric paradigm.* Preager.

Smith, J. W., Anderson, D. H., & Moore, R. L. (2012). Social capital, place meanings, and perceived resilience to climate change. *Rural Sociology, 77*(3), 380–407. https://doi.org/10.1111/j.1549-0831.2012.00082.x

Snel, E., el Farisi, B., Engbersen, G., & Krouwel, A. (2022). Sociaaleconomische status en institutioneel vertrouwen in een tijd van corona. *Tijdschrift Sociologie, 3.*

Snel, E., Engbersen, G., Boom, G. E. J. de, & Seidler, Y. (2024). *Nederland na de pandemie: Over de doorwerking van de coronapandemie op sociale cohesie.* https://www.risbo.nl/media/119293

Snel, E., Engbersen, G., de Boom, J., Seidler, Y., Sam, N., Enneking, G., l'Istelle, F., van Heck, L., & van Linder, J. (2023). *Corona & Cohesie: Dashboard Sociale Impact Corona. Verslag meting 5 (juni 2022) en meting 6 (November 2022).* Risbo & Erasmus School of Social and Behavioural Sciences.

Speakman, J. R. (2018). Obesity and thermoregulation. *Handbook Clinical Neurology, 156*, 431–443. https://doi.org/10.1016/b978-0-444-63912-7.00026-6

Springvloet, L., Kuipers, M., & Van Laar, M. 2017. *Effecten van stoppen-met- roken ondersteuning onder rokers met een lage sociaal economische status.* https://www.trimbos.nl/aanbod/webwinkel/af1559-effecten-van-stoppen-met-roken-ondersteuning-onder-rokers-met-een-lage-sociaal-economische-status/

Stark, A. (2018). *Public inquiries, policy learning, and the threat of future crises.* Oxford University Press.

Steg, L. (2023). Psychology of climate change. *Annual Review of Psychology, 74*, 391–421. https://doi.org/10.1146/annurev-psych-032720-042905

Stel, M., Ketelaar, D., Gutteling, J., Giebels, E., & Kerstholt, J. (2017). *Het alarmeren en informeren van kwetsbare groepen bij crisissituaties.* Universiteit Twente, Vakgroep Psychologie van Conflict, Risico en Veiligheid.

Storymaps. (n.d.). *Ruimtelijk afwegingskader klimaatadaptieve gebouwde omgeving.* https://storymaps.arcgis.com/stories/e14fe0614cc1440496b90ae03e7a2ce0

Stuurgroep Lob van Gennep. (2020). *Notitie Reikwijdte en Detailniveau Lob van Gennep. Te onderzoeken alternatieven en effecten.* https://www.lobvangennep.nl/applications/lobvangennep/files/Downloads/nrd/NRD_LvG_20200129_definitief.pdf

Sweco Nederland, TwynstraGudde, & Decisio. (2024). *Evaluatie wijzigingen ten aanzien van waterveiligheid in de Waterwet in 2014 en 2017, 2024.* https://open.overheid.nl/documenten/dpc-4f9ae16da1f6bfccd46faa746427fb14a11f4de8/pdf

Sweco Nederland. (2024). *Tekort aan groen in Nederlandse steden.* https://www.sweco.nl/wp-content/uploads/sites/5/2024/12/Versteningsonderzoek-NM-Sweco-2024.pdf

Szreter, S. (2002). The state of social capital: Bringing back in power, politics, and history. *Theory and Society, 31*(5), 573–621. http://www.jstor.org/stable/3108542

Tavory, I., & Wagner-Pacifici, R. (2022). Climate change as an event. *Poetics, 93*, Article 101600. https://doi.org/10.1016/j.poetic.2021.101600

Ten Velden, C., Visser, M., & Van der Strate, E. (2022). NKWK-KBS *Monitoring Lokale Klimaatbestendigheid, Fase 2: Pilot indicatoren voor lokale klimaatbestendigheid op basis van bestaande datasets.* Deltares. https://www.deltares.nl/en/expertise/publications/rapportage-nkwk-kbs-inventarisatie-monitoring-lokale-klimaatbestendigheid-fase-1

Tenbrink, T., & Willcock, S. (2023). Place attachment and perception of climate change as a threat in rural and urban areas. *PLoS ONE, 18*(9), Article e0290354.

Terhoeve, M. (2023). *Ode aan de moddermeisjes.* https://www.eilandennieuws.nl/watersnoodramp/watersnoodramp/14465/ode-aan-de-moddermeisjes

Terpstra, T. (2010). *Flood preparedness: Thoughts, feelings and intentions of the Dutch public.* University of Twente. https://doi.org/10.3990/1.9789036529549

Terpstra, T. (2011). Emotions, Trust, and Perceived Risk: Affective and Cognitive Routes to Flood Preparedness Behavior. *Risk Analysis, 31*(10), 1658–1675. https://doi.org/10.1111/j.1539-6924.2011.01616.x

Tezar, T., & Setiadi, R. (2023). Risk perception of small Islands community on climate change: Evidence From Mepar and Baran Islands Indonesia. *Island Studies Journal, 19*(1), 1–21.

Thaler, R. H., & Shefrin, H. M. (1981). An Economic Theory of Self-Control. *Journal of Political Economy, 89*(2), 392–406. http://www.jstor.org/stable/1833317

Thomalla, F., Smith, R., & Schipper, E. L. F. (2015). Cultural Aspects of Risk to Environmental Changes and Hazards. In *Disaster's impact on livelihood and cultural survival: Losses, opportunities, and mitigation*, Routledge.

Thomas, K., Hardy, R. D., Lazrus, H., Mendez, M., Orlove, B., Rivera-Collazo, I., Roberts, J. T., Rockman, M., Warner, B. P., & Winthrop, R. (2018). Explaining differential vulnerability to climate change: A social science review. *Wires Climate Change, 10*(2), Article e565. https://doi.org/10.1002/wcc.565

Thomas, W. I. (1928). The behavior pattern and the situation. *Publications of the American Sociological Society, 22*, 1–14.

Thoresen, S., Birkeland, M., Wentzel-Larsen, T., & Blix, I. (2018). Loss of trust may never heal. institutional trust in disaster victims in a long-term perspective: Associations With social support and mental health. *Frontiers in Psychology, 9*, 1–10. https://doi.org/10.3389/fpsyg.2018.01204

Tiemeijer, W. (2017). *Wat is er mis met maatschappelijke scheidslijnen?* Wetenschappelijke Raad voor het Regeringsbeleid. Verkenning nr.35.

TNO. (2023a). *Energiearmoede in Nederland 2022.* https://www.rijksoverheid.nl/documenten/rapporten/2023/01/27/tno-rapport-energiearmoede-in-nederland-2022

TNO. (2023b). *Gezondheidskosten en energiearmoede: Een empirische analyse voor Nederland.* https://publications.tno.nl/publication/34640440/YyUGjI/TNO-2023-gezondheidsrisico.pdf

TNO. (n.d.). *Factsheet Climate Proof Cities: Hittestress.* https://www.tno.nl/media/3959/factsheet-hittestress.pdf

Townshend, I., Awosoga, O., Kulig, J., & Fan, H. (2014). Social cohesion and resilience across communities that have experienced a disaster. *Natural Hazards, 76*(2), 913–938. Available on https://doi.org/10.1007/s11069-014-1526-4

Tradowsky, J. S., Philip, S. Y., Kreienkamp, F., Kew, S. F., Lorenz, P., Arrighi, J., Bettmann, T., Caluwaerts, S., Chan, S. C., & De Cruz, L. (2023). Attribution of the heavy rainfall events leading to severe flooding in Western Europe during July 2021. *Climatic Change, 176*(7), 1–38.

U.S. House of Representatives Testimony. (1980). *Federal Efforts to Aid Low-income and Elderly Individuals Affected by Life-Threatening Heat Conditions.* U.S. Government Printing Office.

UNICEF. (2024). *Beat the heat: Child health amid heatwaves in Europe and Central Asia.* https://www.unicef.org/eca/reports/beat-heat-2024

Universitat Autònoma de Barcelona. (n.d.). *Beschrijving case study Itteren en Borharen in de gemeente Maastricht.* https://webs.uab.cat/capflo/case-studies/the-netherlands-dutch/

Universiteit van Nederland. (2022). Videocollege: waarom zijn mensen in de zomer agressiever? *GGZ Nieuws.* https://www.ggznieuws.nl/videocollege-waarom-zijn-mensen-in-de-zomer-agressiever/

Van Buuren, A., & Teisman, G. (2014). *Samen verder werken aan de Delta: de governance van het Deltaprogramma na 2014.* Erasmus Universiteit Rotterdam.

Van Buuren, A. (2019). *The dutch delta approach: The successful reinvention of a policy success.* Oxford University Press. https://doi.org/10.1093/oso/9780198843719.003.0011

Van den Adel, I., & Bekker, B. (2023). *Daisy heeft last van hittestress.* Argos. https://www.vpro.nl/argos/lees/onderwerpen/thermo-staat/2023/daisy-heeft-last-van-hittestress.html

Van den Berg, K., Kolen, B., & Rudolph, M. (2022). *Evacuatiegedrag van getrof- fenen tijdens de overstromingen in Limburg in juli 2021* (PR3591.60). HKV Lijn in water. https://open.rijkswaterstaat.nl/open-overheid/onderzoeksrapporten/@257015/evacuatiegedrag-getroffenen-tijdens/

Van den Brink, M. (2021). Rijkswaterstaat: Guardian of the Dutch delta. In A. Boin, L. A. Fahy, & P. 't Hart (Eds.), *Guardians of public value: How public organisations become and remain institutions* (pp. 237–261).

Van den Broek, M., Harms, J., & Smit, F. (2020). *Gedragsaanpak: Financiële prikkels voor klimaatadaptieve maatregelen.* Behavioural Insights Nederland/Inspire to Act. https://klimaatadaptatienederland.nl/@240732/gedragsaanpak-financiele-prikkels/

Van der Laan, L. N., van der Waal, N. E., & De Wit, J. (2022). *Eindrapportage Corona Melder Evaluatie-Survey LISS panel wave 6.* Tilburg University.

Van der Linden, S. (2017). *Determinants and measurement of climate change risk perception, worry, and concern.* Oxford University Press.

Van der Sanden, E. (2019). Anderhalf jaar in Nederland na orkaan Irma: een leven met financiële problemen, vooroordelen en trauma's. *Caribisch Netwerk.* https://land-na-orkaan-irma-een-leven-met-financiele-problemen-vooroordelen-en-traumas/

Van der Steen, M., Scherpenisse, J., De Jong, I., Boonsta, F., & Arnouts, R. (2016). *De volgende stap. Sturen met maatschappelijke energie in het natuurdomein.* Nederlandse School voor Openbaar Bestuur (NSOB).

Van Duin, M. (1992). *Van rampen leren: een vergelijkend onderzoek naar de lessen uit spoorwegongevallen, hotelbranden en industriële ongelukken.* Haagse Drukkerij.

Van Duin, M. (2011). *Veerkrachtige crisisbeheersing: nuchter over het bijzondere.* Nationaal Instituut Publieke Veiligheid. https://nipv.nl/wp-content/uploads/2023/01/20110201-NIFV-PA-Veerkrachtige-crisisbeheersing-nuchter-over-het-bijzondere-lectorale-rede-M-van-Duin.pdf

Van Holm, E. J., & Wyczalkowski, C. K. (2018). Gentrification in the wake of a hurricane: New Orleans after Katrina. *Urban Studies, 56*(13), 2763–2778. https://doi.org/10.1177/0042098018800445

Van Houwelingen, P. (2016). Indicatoren voor sociale cohesie: Het speciale geval van Japan en zijn implicaties. *Mens & Maatschappij, 91*(2), 153–171.

Van Maanen, L., Van der Mijn, R., Van Beurden, M. H. P. H., Roijendijk, L. M. M., Kingma, B. R. M., Mileti, S., & Van Rijn, H. (2019). Core body temperature speeds up temporal processing and choice behavior under deadlines. *Scientific Reports, 9*(1), 10053. https://doi.org/10.1038/s41598-019-46073-3

Van Mersbergen, S. (2021). *Half miljard euro schade in Limburg, opruimen kan wel vier maanden duren*. https://www.ad.nl/binnenland/half-miljard-euro-schade-in-limburg-opruimen-kan-wel-vier-maanden-duren~a1db2afa/

Van Merwijk, K. (2023). *Verkoeling bij hitte* Academische Werkplaats Gezonde Leefomgeving. https://awgl.nl/projecten/verkoeling-bij-hitte

Van Popering-Verkerk, J., Teisman, G. & Van Buuren, A. (2019). *Eb en vloed van het Deltaprogramma: De Deltacommunity anno 2019 in beeld gebracht*. Erasmus University Rotterdam.

Van Thiel, L., & Mol, P. (2020). *Waterbewustzijn van de Nederlandse bevolking*. In opdracht van Kantar en HWBP. https://klimaatadaptatienederland.nl/@237838/waterbewustzijn-van-de-nederlandse-bevolking/

Van Valkengoed, A. M., & Steg, L. (2019a). Meta-analyses of factors motivating climate change adaptation behaviour. *Nature Climate Change, 9*(2), 158–163.

Van Valkengoed, A. M., & Steg, L. (2019b). *The psychology of climate change adaptation*. Cambridge University Press.

Van Valkengoed, A. M., Perlaviciute, G., & Steg, L. (2022). Relationships between climate change perceptions and climate adaptation actions: Policy support, information seeking, and behaviour. *Climatic Change, 171*(1), 14. https://doi.org/10.1007/s10584-022-03338-7

Van Valkengoed, A. M., Perlaviciute, G., & Steg, L. (2024). From believing in climate change to adapting to climate change: The role of risk perception and efficacy beliefs. *Risk Analysis, 44*(3), 553–565. https://doi.org/10.1111/risa.14193

Van Valkengoed, A., Kreuger, M., Geerts, B., & Heessels, M. (2021). *Vervolgonderzoek Subsidie voor klimaatadaptatie*. Rijksuniversiteit Groningen.

Van Valkengoed, A., Kreuger, M., Peters, A., Geerts, B., & Heessels, M. (2020). *Subsidies voor klimaatadaptatie*. Gemeente Breda.

Van Westen, R. M., & Dijkstra, H. A. (2023). Asymmetry of AMOC Hysteresis in a State-of-The-Art Global Climate Model. *Geophysical Research Letters, 50*(22), e2023GL106088.

Van Westen, R. M., Kliphuis, M., & Dijkstra, H. A. (2024). Physics-based early warning signal shows that AMOC is on tipping course. *Science Advances, 10*(6), eadk1189. https://doi.org/10.1126/sciadv.adk1189

Veiligheidsregio Limburg-Noord, & Veiligheidsregio Zuid-Limburg. (2020). *Rampbestrijdingsplan Hoogwater Maas Limburg 2020–2023*. https://lokaleregelgeving.overheid.nl/CVDR719417/1

VenW. (2009). *Nationaal Waterplan 2009–2015*. Ministerie van Verkeer en Waterstaat. https://zoek.officielebekendmakingen.nl/blg-47417.pdf

Verbond van Verzekeraars. (2022). *Daan Prevoo, burgemeester van Valkenburg*. https://www.verzekeraars.nl/publicaties/longreads/daan-prevoo-burgemeester-van-valkenburg

Verenigde Naties (VN). (2015). *Paris Agreement. United Nations treaty collect*, 1–27. https://unfccc.int/sites/default/files/english_paris_agreement.pdf

Vermeulen, S. J., Campbell, B. M., & Ingram, J. S. I. (2012). Climate Change and Food Systems. *Annual Review of Environment and Resources, 37*, 195–222. https://doi.org/10.1146/annurev-environ-020411-130608

Verordening (EU). (2021). 2021/1119 van het Europees Parlement en de Raad van 30 juni 2021 tot vaststelling van een kader voor de verwezenlijking van klimaatneutraliteit, en tot wijziging van Verordening (EG) nr. 401/2009 en Verordening (EU) 2018/1999 ("Europese klimaatwet"), 1–17. https://climate.ec.europa.eu/eu-action/european-climate-law_nl

Visser, H., Van Prooije, H., De Vries, H., & Petersen, A.C. (2025). The likelihood of holding outdoor skating marathons: the past, present and future of a climate-change indicator, and a way to adapt. *Climate Change, 178*. https://doi.org/10.1007/s10584-025-03920-9

VZinfo. (n.d.). *Wat is sociaaleconomische status?* https://www.vzinfo.nl/sociaaleconomische-status/verantwoording/definities

Wall, T. (2019, mei 18). 'This is a wake-up call': the villagers who could be Brittain's first climate refugees. *The Guardian*. https://www.theguardian.com/environment/2019/may/18/this-is-a-wake-up-call-the-villagers-who-could-be-britains-first-climate-refugees

Wargocki, P., Porras-Salazar, J. A., & Contreras-Espinoza, S. (2019). The relationship between classroom temperature and children's performance in school. *Building and Environment, 157*, 197–204. https://doi.org/10.1016/j.buildenv.2019.04.046

Werkgroep Dorpsraden. (2024). *Kunnen de gevolgen van een overstroming in de Lob van Gennep worden beperkt?* https://maasburen.nl/files/maasburen.nl/uploads/files/740811/Nota%20Overstromingsbestendig%20maken%20Lob%20v%20G%20jan2024.pdf

Westerhof, S., Taanman, M., Van Popering-Verkerk, J., & Duijn, M. (2023). *Why Inequality matters to Resilience*. Erasmus Universiteit Rotterdam. https://convergence.nl/app/uploads/RDI_veerkracht-en-ongelijkheid_V3.pdf

Whitney, J. (2021). *A new mantra for Japanese disaster management*. https://eastasiaforum.org/wp-content/uploads/2023/12/V.13-N.3.pdf

Wijkhuis et al. (2018). Orkaan Irma treft Sint Maarten en Caribisch Nederland. Een evaluatie van het doorhet ministerie van IenW geleverde crisismanagement. IFV Instituut Fysieke Veiligheid.

Wildavsky, A. (1988). Searching for safety. *Routledge*. https://doi.org/10.4324/9781351316248

Williams, M. N., Hill, S. R., & Spicer, J. (2015). The relationship between temperature and assault in New Zealand. *Climatic Change, 132*(4), 559–573. https://doi.org/10.1007/s10584-015-1438-7

Witte, J. P. M., De Louw, P. G., van Ek, R., Bartholomeus, R. P., Van den Eertwegh, G., Gilissen, H. K., Beugelink, G., Ruijtenberg, R., & Van der Kooij, W. (2020). Aanpak droogte vraagt transitie waterbeheer. *Water Governance, 3*, 120–131.

Witte, K. (1992). Putting the fear back into fear appeals: The extended parallel process model. *Communication Monographs, 59*(4), 329–349. https://doi.org/10.1080/03637759209376276

WKR. (2023). *Met iedereen de transities in Richtinggevende keuzes voor een klimaatneutraal en klimaatbestendig Nederland*. https://www.wkr.nl/documenten/rapporten/2023/12/15/adviesrapport-met-iedereen-de-transities-in

WRR. (2008). *Onzekere veiligheid* (rapport nr.82). Wetenschappelijke Raad voor het Regeringsbeleid. https://www.wrr.nl/publicaties/rapporten/2008/10/01/onzekere-veiligheid

WRR. (2014). *Met kennis van gedrag beleid maken* (rapport nr.92). Wetenschappelijke Raad voor het Regeringsbeleid. https://www.wrr.nl/publicaties/rapporten/2014/09/10/met-kennis-van-gedrag-beleid-maken

WRR. (2017a). *Wat is er mis met maatschappelijke scheidslijnen?* (verkenningen nr.35). Wetenschappelijke Raad voor het Regeringsbeleid. https://www.wrr.nl/publicaties/verkenningen/2017/01/12/wat-is-er-mis-met-maatschappelijke-scheidslijnen---35

WRR. (2017b). *Weten is nog geen doen. Een realistisch perspectief op redzaamheid* (rapport nr.97). Wetenschappelijke Raad voor het Regeringsbeleid. https://www.wrr.nl/publicaties/rapporten/2017/04/24/weten-is-nog-geen-doen

WRR. (2018a). *De nieuwe verscheidenheid. Toenemende diversiteit naar herkomst in Nederland* (verkenningen nr.38). Wetenschappelijke Raad voor het Regeringsbeleid. https://www.wrr.nl/publicaties/verkenningen/2018/05/29/de-nieuwe-verscheidenheid

WRR. (2018b). *Van verschil naar potentieel. Een realistisch perspectief op de sociaaleconomische gezondheidsverschillen* (policy brief). Wetenschappelijke Raad voor het Regeringsbeleid. https://www.wrr.nl/publicaties/policy-briefs/2018/08/27/van-verschil-naar-potentieel.-een-realistisch-perspectief-op-de-sociaaleconomische-gezondheidsverschillen

WRR. (2019). *Voorbereiden op digitale ontwrichting* (rapport nr.101). Wetenschappelijke Raad voor het Regeringsbeleid. https://www.wrr.nl/adviesprojecten/digitale-ontwrichting

WRR. (2022). *Coronascenario's doordacht: Handreiking voor noodzakelijke keuzes*. Wetenschappelijke Raad voor het Regeringsbeleid. https://www.wrr.nl/publicaties/publicaties/2022/09/05/coronascenarios-doordacht-handreiking-voor-noodzakelijke-keuzes

WRR. (2023a). *Rechtvaardigheid in klimaatbeleid: Over de verdeling van klimaatkosten* (rapport nr.106). Wetenschappelijke Raad voor het Regeringsbeleid. https://www.wrr.nl/adviesprojecten/klimaatbeleid/documenten/rapporten/2023/02/16/rechtvaardigheid-in-klimaatbeleid

WRR. (2023b). *Goede zaken: Naar een grotere maatschappelijke bijdrage van ondernemingen* (rapport nr.107). Wetenschappelijke Raad voor het Regeringsbeleid. https://www.wrr.nl/publicaties/rapporten/2023/09/14/goede-zaken

WRR. (2023c). *Grip: Het maatschappelijk belang van persoonlijke controle* (rapport nr.108). Wetenschappelijke Raad voor het Regeringsbeleid. https://www.wrr.nl/publicaties/rapporten/2023/11/30/grip

WRR. (2024). *Nederland in een fragmenterende wereldorde* (rapport nr.109). Wetenschappelijke Raad voor het Regeringsbeleid. https://www.wrr.nl/adviesprojecten/kantelende-wereldorde/documenten/rapporten/2024/07/01/nederland-in-een-fragmenterende-wereldorde

Xu, C., Kohler, T. A., Lenton, T. M., Svenning, J. -C., & Scheffer, M. (2020). Future of the human climate niche. *Proceedings of the National Academy of Sciences, 117*(21), 11350–11355. https://doi.org/10.1073/pnas.1910114117

Zamboni, L. M., & Martin, E. G. (2020). Association of US households' disaster preparedness with socioeconomic characteristics, composition, and region. *JAMA Network Open, 3*(4), Article e206881. https://doi.org/10.1001/jamanetworkopen.2020.6881

Zhang, F., De Dear, R., & Hancock, P. (2019). Effects of moderate thermal environments on cognitive performance: A multidisciplinary review. *Applied Energy, 236*, 760–777. https://doi.org/10.1016/j.apenergy.2018.12.005

Zscheischler, J., Westra, S., Van den Hurk, B. J. J. M., Seneviratne, S. I., Ward, P. J., Pitman, A., Agha Kouchak, A., Bresch, D. N., Leonard, M., Wahl, T., & Zhang, X. (2018). Future climate risk from compound events. *Nature Climate Change, 8*(6), 469–477. https://doi.org/10.1038/s41558-018-0156-3

Zylberman, P. (2012). Crises sanitaires, crises politiques. *Les Tribunes De La Santé, 34*(1), 35–50. https://doi.org/10.3917/seve.034.0035

Index

R. El-Dardiry et al., *People and Climate*, Research for Policy,
https://doi.org/10.1007/978-3-032-25831-1

N

O

P

R

S

T

U

V

W

FSC
www.fsc.org
MIX
Papier | Fördert
gute Waldnutzung
FSC® C083411

Kolibri 360 GmbH
Ferdinand-Jühlke-Straße 7
99095 Erfurt, Deutschland
produktsicherheit@kolibri360.de